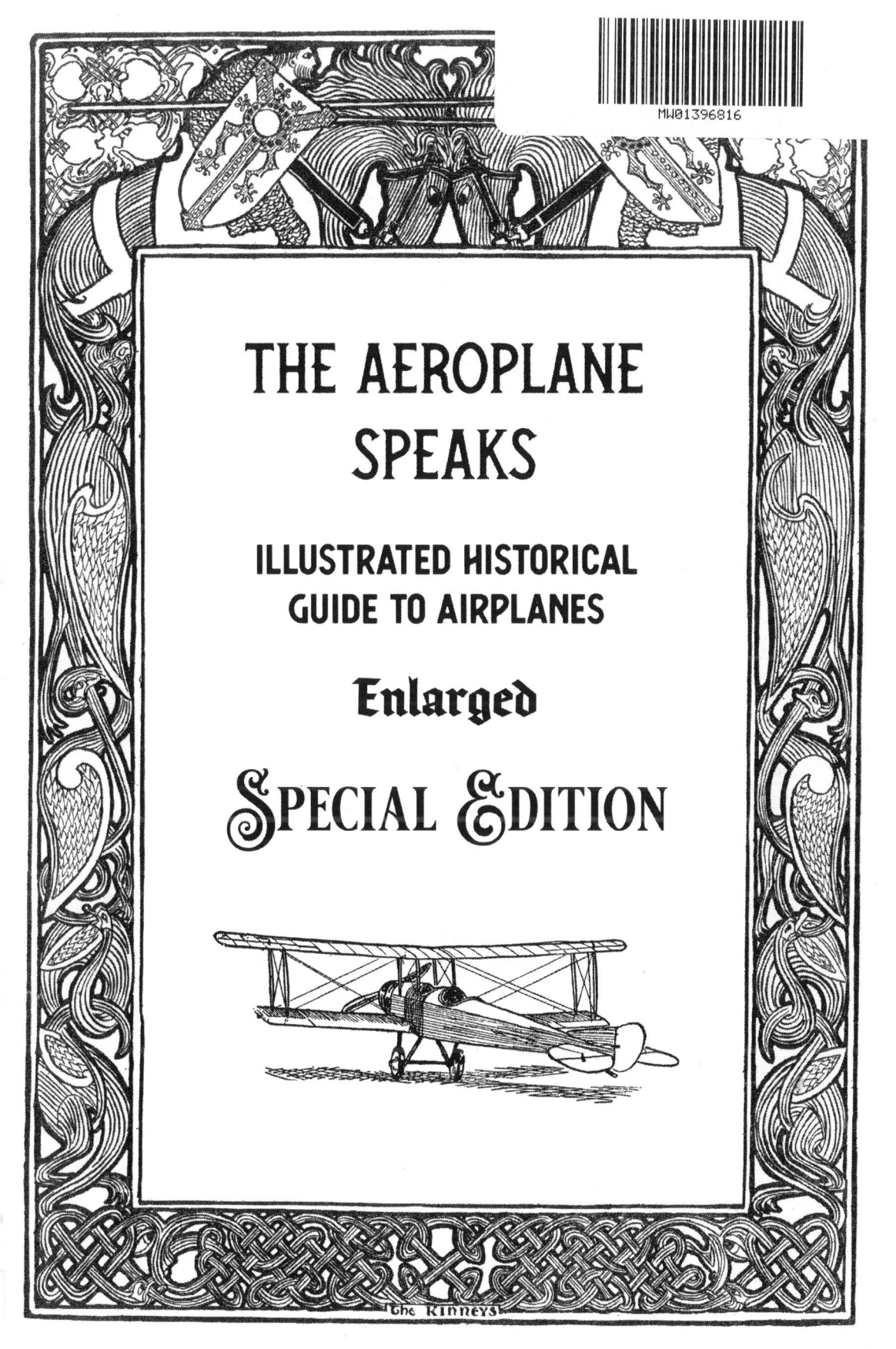

THE AEROPLANE SPEAKS

ILLUSTRATED HISTORICAL GUIDE TO AIRPLANES

Enlarged

Special Edition

The Aeroplane Speaks:
Illustrated Historical Guide to Airplanes
Enlarged Special Edition

by H. Barber

Copyright © 2020 Inecom, LLC.
All Rights Reserved

Cover Design by
Mark Bussler

More History Books at
CGRpublishing.com

1939 New York World's Fair: The World of Tomorrow in Photographs

The Art Of World War 1

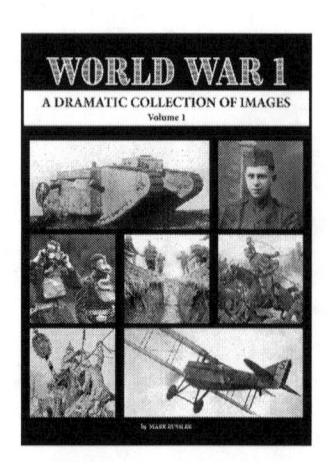

World War 1: A Dramatic Collection of Images

CONTENTS

PROLOGUE

	PAGE
PART I.—THE ELEMENTARY PRINCIPLES AIR THEIR GRIEVANCES	7
" II.—THE PRINCIPLES, HAVING SETTLED THEIR DIFFERENCES, FINISH THE JOB	21
" III.—THE GREAT TEST	33
" IV.—CROSS COUNTRY	44

CHAPTER I.—FLIGHT	61
" II.—STABILITY AND CONTROL	76
" III.—RIGGING	96
" IV.—PROPELLERS	121
" V.—MAINTENANCE	132
GLOSSARY	139
TYPES OF AEROPLANES	151

High Aspect Ratio
Low Aspect Ratio The Inventor
The Pilot
Lateral Dihedral
Directional Stability Longitudinal Dihedral
Efficiency

Thrust Drift Lift Grandfather Centrifugal Force Stagger
The Designer Gravity
Flight

THE FLIGHT FOLK.

MOTIVE

THE reasons impelling me to write this book, the maiden effort of my pen, are, firstly, a strong desire to help the ordinary man to understand the Aeroplane and the joys and troubles of its Pilot; and, secondly, to produce something of *practical* assistance to the Pilot and his invaluable assistant the Rigger. Having had some eight years' experience in designing, building, and flying aeroplanes, I have hopes that the practical knowledge I have gained may offset the disadvantage of a hand more used to managing the "joystick" than the dreadful haltings, the many side-slips, the irregular speed, and, in short, the altogether disconcerting ways of a pen.

The matter contained in the Prologue appeared in the *Field* of May 6th, 13th, 20th, and 27th, 1916, and is now reprinted by the kind permission of the editor, Sir Theodore Cook.

I have much pleasure in also acknowledging the kindness of Mr. C. G. Grey, editor of the *Aeroplane,* to whom I am indebted for the valuable illustrations reproduced at the end of this book.

DEDICATED

TO THE

SUBALTERN FLYING OFFICER

THE AEROPLANE SPEAKS

PROLOGUE

PART I

THE ELEMENTARY PRINCIPLES AIR THEIR GRIEVANCES

THE Lecture Hall at the Royal Flying Corps School for Officers was deserted. The pupils had dispersed, and the Officer Instructor, more fagged than any pupil, was out on the aerodrome watching the test of a new machine.

Deserted, did I say? But not so. The lecture that day had been upon the Elementary Principles of Flight, and they lingered yet. Upon the Blackboard was the illustration you see in the frontispiece.

"I am the side view of a Surface," it said, mimicking the tones of the lecturer. "Flight is secured by driving me through the air at an angle inclined to the direction of motion."

"Quite right," said the Angle. "That's me, and I'm the famous Angle of Incidence."

"And," continued the Surface, "my action is to deflect the air downwards, and also, by fleeing from the air behind, to create a semi-vacuum or rarefied area over most of the top of my surface."

"This is where I come in," a thick, gruff voice was heard, and went on: "I'm the Reaction. You can't have action without me. I'm a very considerable force, and my direction is at right-angles to you," and he looked heavily at the Surface. "Like this," said he, picking up the chalk with his Lift, and drifting to the Blackboard.

"I act in the direction of the arrow R, that is, more or less, for the direction varies somewhat with the Angle of Incidence and the curvature of the Surface; and, strange but true, I'm stronger on the top of the Surface than at the bottom of it. The Wind Tunnel has proved that by exhaustive research—and don't forget how quickly I can grow! As the speed through the air increases my strength increases more rapidly than you might think—approximately, as the Square of the Speed; so you see that if the Speed of the Surface through the air is, for instance, doubled, then I am a good deal more than doubled. That's because I am the result of not only the mass of air displaced, but also the result of the Speed with which the Surface engages the Air. I am a product of those two factors, and at the speeds at which Aeroplanes fly to-day, and at the altitudes

The action of the surface upon the air.

and consequent density of air they at present experience, I increase at about the Square of the Speed.

"Oh, I'm a most complex and interesting personality, I assure you—in fact, a dual personality, a sort of aeronautical Dr. Jekyll and Mr. Hyde. There's Lift, my vertical part or *component,* as those who prefer long words would say; he always acts vertically upwards, and hates Gravity like poison. He's the useful and admirable part of me. Then there's Drift, my horizontal component, sometimes, though rather erroneously, called Head Resistance; he's a villain of the deepest dye, and must be overcome before flight can be secured."

THE ELEMENTARY PRINCIPLES' GRIEVANCES

"And I," said the Propeller, "I screw through the air and produce the Thrust. I thrust the Aeroplane through the air and overcome the Drift; and the Lift increases with the Speed, and when it equals the Gravity of Weight, then—there you are—Flight! And nothing mysterious about it at all."

"I hope you'll excuse me interrupting," said a very beautiful young lady, "my name is Efficiency, and, while no doubt, all you have said is quite true, and that, as my young man the Designer says, 'You can make a tea-tray

fly if you slap on Power enough,' I can assure you that I'm not to be won quite so easily."

"Well," eagerly replied the Lift and the Thrust, "let's be friends. Do tell us what we can do to help you to overcome Gravity and Drift with the least possible Power. That obviously seems the game to play, for more Power means heavier engines, and that in a way plays into the hands of our enemy, Gravity, besides necessitating a larger Surface or Angle to lift the Weight, and that increases the Drift."

"Very well," from Efficiency, "I'll do my best, though I'm so shy, and I've just had such a bad time at the Factory, and I'm terribly afraid you'll find it awfully dry."

THE AEROPLANE SPEAKS

"Buck up, old dear!" This from several new-comers, who had just appeared. "We'll help you," and one of them, so lean and long that he took up the whole height of the lecture room, introduced himself.

"I'm the High Aspect Ratio," he said, "and what we have got to do to help this young lady is to improve the proportion of Lift to Drift. The more Lift we can get for a certain area of Surface, the greater the Weight the latter can carry; and the less the Drift, then the less Thrust and Power required to overcome it. Now it is a fact that, if

LOW ASPECT RATIO. HIGH ASPECT RATIO.

the Surface is shaped to have the greatest possible span, *i.e.*, distance from wing-tip to wing-tip, it then engages more air and produces both a maximum Reaction and a better proportion of Lift to Drift.

"That being so, we can then well afford to lose a little Reaction by reducing the Angle of Incidence to a degree giving a still better proportion of Lift to Drift than would otherwise be the case; for you must understand that the Lift-Drift Ratio depends very much upon the size of the Angle of Incidence, which should be as small as possible within certain limits. So what I say is, make the surface of

THE ELEMENTARY PRINCIPLES' GRIEVANCES

Infinite Span with no width or *chord,* as they call it. That's all I require, I assure you, to make me quite perfect and of infinite service to Miss Efficiency."

"That's not practical politics," said the Surface. "The way you talk one would think you were drawing £400 a year at Westminster, and working up a reputation as an Aeronautical Expert. I must have some depth and chord to take my Spars and Ribs, and again, I must have a certain chord to make it possible for my Camber (that's curvature) to be just right for the Angle of Incidence. If that's not right the air won't get a nice uniform compression and downward acceleration from my underside, and the rarefied 'suction' area over the top of me will not be as even and clean

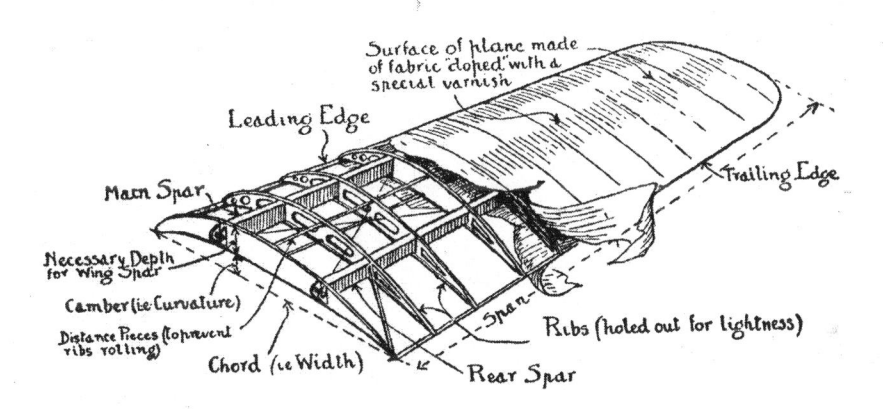

in effect as it might be. That would spoil the Lift-Drift Ratio more than you can help it. Just thrust that chalk along, will you? and the Blackboard will show you what I mean."

"Well," said the Aspect Ratio, "have it your own way, though I'm sorry to see a pretty young lady like Efficiency compromised so early in the game."

"Look here," exclaimed a number of Struts, "we have got a brilliant idea for improving the Aspect Ratio," and with that they hopped up on to the Spars. "Now," excitedly, "place another Surface on top of us. Now do you see? There is double the Surface, and that being so, the proportion of Weight to Surface area is halved. That's less burden of work for the Surface, and so the Spars need not be so strong and so deep, which results in not so thick a Surface. That means the Chord can be proportionately decreased without adversely affecting the Camber. With the Chord decreased, the Span becomes relatively greater, and so produces a splendid Aspect Ratio, and an excellent proportion of Lift to Drift."

"I don't deny that they have rather got me there," said the Drift, "but all the same, don't forget my increase due to the drift of the Struts and their bracing wires."

"Yes, I dare say," replied the Surface, "but remember that my Spars are less deep than before, and consequently I am not so thick now, and shall for that reason also be able to go through the air with a less proportion of Drift to Lift."

"Remember me also, please," croaked the Angle of Incidence. "Since the Surface has now less weight to carry for its area, I may be set at a still lesser and finer Angle. That means less Drift again. We are certainly getting on splendidly! Show us how it looks now, Blackboard." And the Blackboard obligingly showed them as follows:

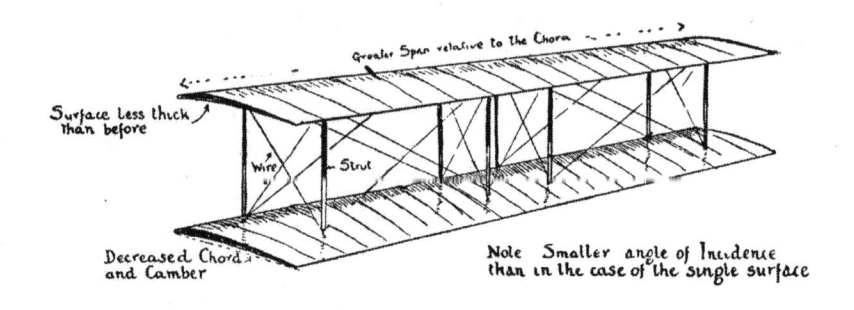

"Well, what do you think of that?" they all cried to the Drift.

"You think you are very clever," sneered the Drift. "But you are not helping Efficiency as much as you think. The suction effect on the top of the lower Surface will give a downward motion to the air above it and the result will be that the bottom of the top Surface will not secure as good a Reaction from the air as would otherwise be the case, and that means loss of Lift; and you can't help matters by increasing the gap between the surfaces because that means longer Struts and Wires, and that in itself would help me, not to speak of increasing the Weight. You see it's not quite so easy as you thought."

THE ELEMENTARY PRINCIPLES' GRIEVANCES

At this moment a hiccough was heard, and a rather fast and rakish-looking chap, named Stagger, spoke up. "How d'ye do, miss," he said politely to Efficiency, with a side glance out of his wicked old eye. "I'm a bit of a knut, and without the slightest trouble I can easily minimize the disadvantage that old reprobate Drift has been frightening you with. I just stagger the top Surface a bit forward, and no longer is that suction effect dead under it. At the same time I'm sure the top Surface will kindly extend its Span for such distance as its Spars will support it without the aid of Struts. Such extension will be quite useful, as there will be no Surface at all underneath it to interfere with the Reaction above." And the Stagger leaned forward and picked up the Chalk, and this is the picture he drew:

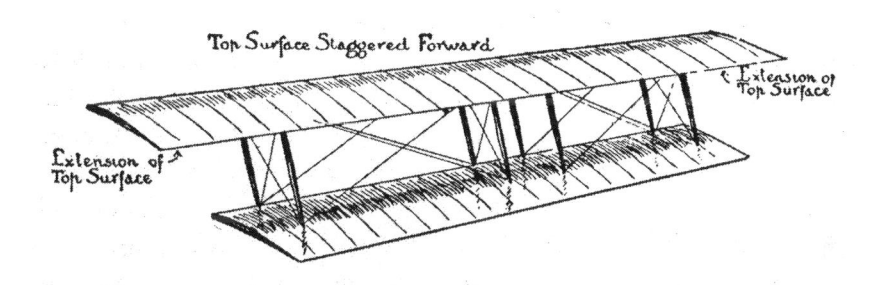

Said the Blackboard, "That's not half bad! It really begins to look something like the real thing, eh?"

"The real thing, is it?" grumbled Drift. "Just consider that contraption in the light of any one Principle, and I warrant you will not find one of them applied to perfection. The whole thing is nothing but a Compromise." And he glared fixedly at poor Efficiency.

"Oh, dear! Oh, dear!" she cried. "I'm always getting into trouble. What *will* the Designer say?"

"Never mind, my dear," said the Lift-Drift Ratio, consolingly. "You are improving rapidly, and quite useful enough now to think of doing a job of work."

"Well, that's good news," and Efficiency wiped her eyes with her Fabric and became almost cheerful. "Suppose we think about finishing it now? There will have to be an Engine and Propeller, won't there? And a body to fix them in, and tanks for oil and petrol, and a tail, and," archly, "one of those dashing young Pilots, what?"

"Well, we are getting within sight of those interesting Factors," said the Lift-Drift Ratio, "but first of all we had better decide upon the Area of the Surfaces, their Angle of Incidence and Camber. If we are to ascend as quickly as possible the Aeroplane must be *slow* in order to secure the best possible Lift-Drift Ratio, for the drift of the struts wires, body, etc., increases approximately as the square of the speed, but it carries with it no lift as it does in the case of the Surface. The less speed then, the less such drift, and the better the Aeroplane's proportion of lift to drift; and, being slow, we shall require a *large Surface* in order to secure a large lift relative to the weight to be carried. We shall also require a *large Angle of Incidence* relative to the horizontal, in order to secure a proper inclination of the Surface to the direction of motion, for you must remember that, while we shall fly upon an even keel and with the propeller thrust horizontal (which is its most efficient attitude), our flight path, which is our direction of motion, will be sloping upwards, and it will therefore be necessary to fix the Surface to the Aeroplane at a very considerable angle relative to the horizontal Propeller Thrust in order to secure a proper angle to the upwards direction of motion. Apart from that, we shall require a larger Angle of Incidence than in the case of a machine designed purely for speed, and that means a correspondingly *large Camber*.

"On the other hand, if we are thinking merely of Speed, then a *Small Surface*, just enough to lift the weight off the ground, will be best, also a *small Angle* to cut the Drift down and that, of course, means a relatively *small Camber*.

"So you see the essentials for *Climb* or quick ascent and for *Speed* are diametrically opposed. Now which is it to be?"

"Nothing but perfection for me," said Efficiency. "What I want is Maximum Climb and Maximum Speed for the Power the Engine produces."

And each Principle fully agreed with her beautiful sentiments, but work together they would not.

The Aspect Ratio wanted infinite Span, and hang the Chord.

THE ELEMENTARY PRINCIPLES' GRIEVANCES

The Angle of Incidence would have two Angles and two Cambers in one, which was manifestly absurd; the Surface insisted upon no thickness whatever, and would not hear of such things as Spars and Ribs; and the Thrust objected to anything at all likely to produce Drift, and very nearly wiped the whole thing off the Blackboard.

There was, indeed, the makings of a very pretty quarrel when the Letter arrived. It was about a mile long, and began to talk at once.

"I'm from the Inventor," he said, and hope rose in the heart of each heated Principle. "It's really absurdly simple.

Maximum climb.

Maximum speed.

All the Pilot has to do is to touch a button, and at his will, VARY the area of the Surface, the Angle of Incidence, and the Camber! And there you are—Maximum Climb or Maximum Speed as required! How does that suit you?"

"That suits us very well," said the Surface, "but, excuse me asking, how is it done without apparatus increasing the Drift and the Weight out of all reason? You won't mind showing us your Calculations, Working Drawings, Stress Diagrams, etc., will you?"

Said the Letter with dignity, "I come from an Inventor

THE AEROPLANE SPEAKS

so brilliantly clever as to be far above the unimportant matters you mention. He is no common working man, sir! He leaves such things to Mechanics. The point is, you press a button and——"

"Look here," said a Strut, rather pointedly, "where do you think you are going, anyway?"

"Well," from the Letter, "as a matter of fact, I'm not addressed yet, but, of course, there's no doubt I shall reach the very highest quarters and absolutely revolutionize Flight when I get there."

Said the Chalk, "I'll address you, if that's all you want; now drift along quickly!" And off went the Letter to The Technical Editor, "Daily Mauler," London.

And a League was formed, and there were Directors with Fees, and several out-of-service Tin Hats, and the Man-who-takes-the-credit, and a fine fat Guinea-pig, and all the rest of them. And the Inventor paid his Tailor and had a Hair-Cut, and is now a recognized *Press* Expert—but he is still waiting for those Mechanics!

"I'm afraid," said the Slide-rule, who had been busy making those lightning-like automatic calculations for which he is so famous, "it's quite impossible to fully satisfy all of you, and it is perfectly plain to me that we shall have to effect a Compromise and sacrifice some of the Lift for Speed."

Thud! What was that?

Efficiency had fainted dead away! The last blow had been too much for her. And the Principles gathered mournfully round, but with the aid of the Propeller Slip* and a friendly lift from the Surface she was at length revived and regained a more normal aspect.

Said the Stagger with a raffish air, "My dear young lady, I assure you that from the experiences of a varied career, I have learned that perfection is impossible, and I am sure the Designer will be quite satisfied if you become the Most Efficient Compromise."

"Well, that sounds so common sense," sighed Efficiency, "I suppose it must be true, and if the Designer is satisfied,

* Propeller Slip: As the propeller screws through the air, the latter to a certain extent gives back to the thrust of the propellor blades, just as the shingle on the beach slips back as you ascend it. Such "give-back" is known as "slip," and anyone behind the propeller will feel the slip as a strong draught of air.

THE ELEMENTARY PRINCIPLES' GRIEVANCES

that's all I really care about. Now do let's get on with the job."

So the Chalk drew a nice long slim body to hold the Engine and the tanks, etc., with room for the Pilot's and Passenger's seats, and placed it exactly in the middle of the Biplane. And he was careful to make its position such that the Centre of Gravity was a little in advance of the Centre of Lift, so that when the Engine was not running and there was consequently no Thrust, the Aeroplane should be "nose-heavy" just to the right degree, and so take up a natural glide to Earth—and this was to help the Pilot and relieve him of work and worry, should he find himself in a fog or

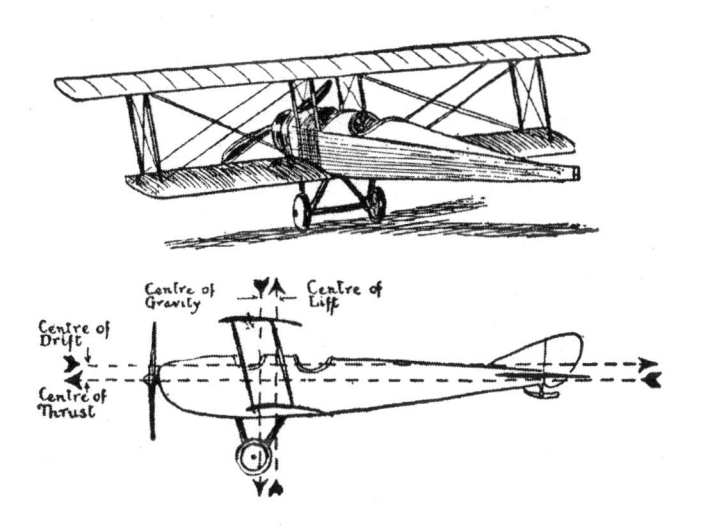

a cloud. And so that this tendency to glide downwards should not be in evidence when the Engine was running and descent not desired, the Thrust was placed a little below the Centre of Drift or Resistance. In this way it would in a measure pull the nose of the Aeroplane up and counterbalance the "nose-heavy" tendency.

And the Engine was so mounted that when the Propeller-Thrust was horizontal, which is its most efficient position, the Angle of Incidence and the Area of the surfaces were just sufficient to give a Lift a little in excess of the Weight. And the Camber was such that, as far as it was concerned, the Lift-Drift Ratio should be the best possible for that Angle of Incidence. And a beautifully simple under-carriage was added, the outstanding features of which were simplicity, strength, light-weight, and minimum drift. And, last of all, there was the Elevator, of which you will hear more by-and-by. And this is what it looked like then:

And Efficiency, smiling, thought that it was not such a bad compromise after all and that the Designer might well be satisfied.

"Now," said she, "there's just one or two points I'm a bit hazy about. It appears that when the Propeller shaft is horizontal and so working in its most efficient attitude, I shall have a Lift from the Surfaces slightly in excess of the Weight. That means I shall ascend slightly, at the same time making nearly maximum speed for the power and thrust. Can't I do better than that?"

"Yes, indeed," spoke up the Propeller, "though it means that I must assume a most undignified attitude, for helicopters* I never approved of. In order to ascend more quickly the Pilot will deflect the Elevator, which, by the way, you see hinged to the Tail. By that means he will force the whole Aeroplane to assume a greater Angle of Incidence. And with greater Angle, the Lift will increase, though I'm sorry to say the Drift will increase also. Owing to the greater Drift, the Speed through the air will lessen, and I'm afraid that won't be helpful to the Lift; but I shall now be pointing upwards, and besides overcoming the Drift in a forward direction I shall be doing my best to haul the Aeroplane skywards. At a certain angle known as the Best Climbing Angle, we shall have our Maximum Margin of Lift, and I'm hoping that may be as much as almost a thousand feet altitude a minute."

*Helicopter. An air-screw revolving upon a vertical axis. If driven with sufficient power, it will lift vertically, but having regard to the mechanical difficulties of such construction, it is a most inefficient way of securing lift compared with the arrangement of an inclined surface driven by a propeller revolving about a horizontal axis.

THE ELEMENTARY PRINCIPLES' GRIEVANCES

"Then, if the Pilot is green, my chance will come," said the Maximum Angle of Incidence. "For if the Angle is increased over the Best Climbing Angle, the Drift will rush up; and the Speed, and with it the Lift, will, when my Angle is reached, drop to a point when the latter will be no more than the Weight. The Margin of Lift will have entirely disappeared, and there we shall be, staggering along at my tremendous angle, and only just maintaining horizontal flight."

"And then with luck I'll get my chance," said the Drift. "If he is a bit worse than green, he'll perhaps still further

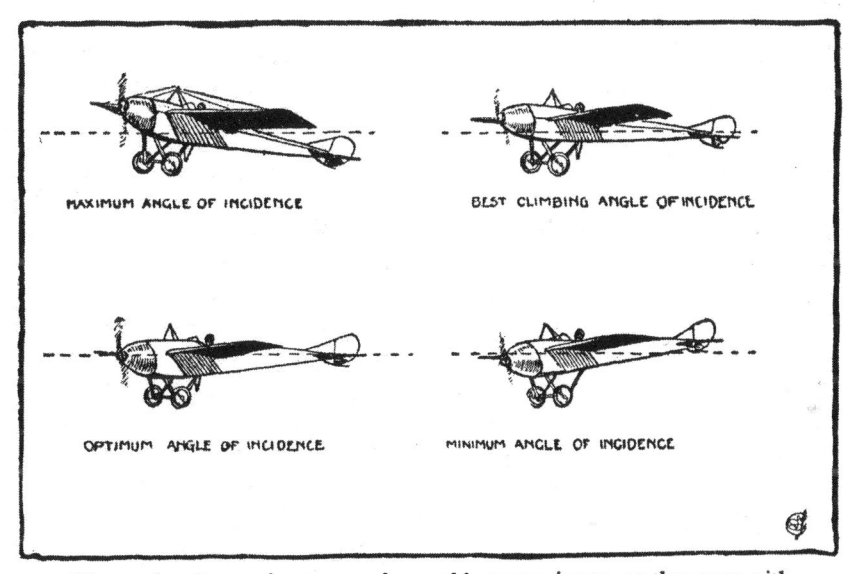

MAXIMUM ANGLE OF INCIDENCE

BEST CLIMBING ANGLE OF INCIDENCE

OPTIMUM ANGLE OF INCIDENCE

MINIMUM ANGLE OF INCIDENCE

The angles shown above are only roughly approximate, as they vary with different types of aeroplanes

increase the Angle. Then the Drift, largely increasing, the Speed, and consequently the Lift, will become still less, *i.e.*, less than the Weight, and then—what price pancakes,* eh?"

"Thank you," from Efficiency, "that was all most informing. And now will you tell me, please, how the greatest Speed may be secured?"

"Certainly, now it's my turn," piped the Minimum Angle of Incidence. "By means of the Elevator, the Pilot places

* Pancakes: Pilot's slang for stalling an aeroplane and dropping like a pancake.

the Aeroplane at my small Angle, at which the Lift only just equals the Weight, and, also, at which we shall make greater speed with no more Drift than before. Then we get our greatest Speed, just maintaining horizontal flight."

"Yes; though I'm out of the horizontal and thrusting downwards," grumbled the Propeller, "and that's not efficient, though I suppose it's the best we can do until that Inventor fellow finds his Mechanics."

"Thank you so much," said Efficiency. "I think I have now at any rate an idea of the Elementary Principles of Flight, and I don't know that I care to delve much deeper, for sums always give me a headache; but isn't there something about Stability and Control? Don't you think I ought to have a glimmering of them too?"

"Well, I should smile," said a spruce Spar, who had come all the way from America. "And that, as the Lecturer says, 'will be the subject of our next lecture,' so be here again to-morrow, and you will be glad to hear that it will be distinctly more lively than the subject we have covered to-day."

PART II

THE PRINCIPLES, HAVING SETTLED THEIR DIFFERENCES, FINISH THE JOB

ANOTHER day had passed, and the Flight Folk had again gathered together and were awaiting the arrival of Efficiency who, as usual, was rather late in making an appearance.

The crowd was larger than ever, and among the newcomers some of the most important were the three Stabilities, named Directional, Longitudinal, and Lateral, with their assistants, the Rudder, Elevator, and Ailerons. There was Centrifugal Force, too, who would not sit still and created a most unfavourable impression, and Keel-Surface, the Dihedral Angle, and several other lesser fry.

"Well," said Centrifugal Force, "I wish this Efficiency I've heard so much about would get a move on. Sitting still doesn't agree with me at all. Motion I believe in. There's nothing like motion—the more the better."

"We are entirely opposed to that," objected the three Stabilities, all in a breath. "Unless it's in a perfectly straight line or a perfect circle. Nothing but perfectly straight lines or, upon occasion, perfect circles satisfy us, and we are strongly suspicious of your tendencies."

"Well, we shall see what we shall see," said the Force darkly. "But who in the name of blue sky is this?"

And in tripped Efficiency, in a beautifully "doped" dress of the latest fashionable shade of khaki-coloured fabric, a perfectly stream-lined bonnet, and a bewitching little Morane parasol,* smiling as usual, and airily exclaiming, "I'm so sorry I'm late, but you see the Designer's such a funny man. He objects to skin friction,† and insisted upon me changing my fabric for one of a smoother

---

* Morane parasol: A type of Morane monoplane in which the lifting surfaces are raised above the pilot in order to afford him a good view of the earth.
† Skin friction is that part of the drift due to the friction of the air with roughnesses upon the surface of the aeroplane.

surface, and that delayed me. Dear me, there are a lot more of us to-day, aren't there? I think I had better meet one at a time." And turning to Directional Stability, she politely asked him what he preferred to do.

"My purpose in life, miss," said he, "is to keep the Aeroplane on its course, and to achieve that there must be, in effect, more Keel-Surface behind the Vertical Turning Axis than there is in front of it."

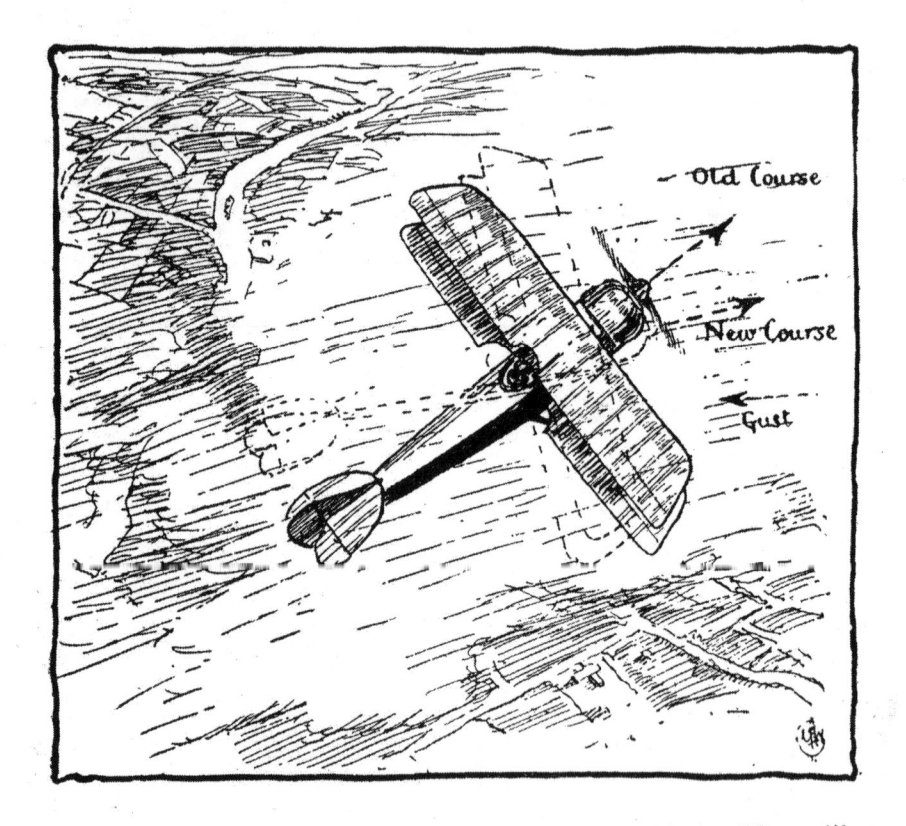

Efficiency looking a little puzzled, he added: "Just like a weathercock, and by Keel-Surface I mean everything you can see when you view the Aeroplane from the side of it—the sides of the body, struts, wires, etc."

"Oh, now I begin to see light," said she: "but just exactly how does it work?"

"I'll answer that," said Momentum. "When perhaps by a gust of air the Aeroplane is blown out of its course and points in another direction, it doesn't immediately fly off on that new course. I'm so strong I pull it off the

THE PRINCIPLES FINISH THE JOB

new course to a certain extent, and towards the direction of the old course. And so it travels, as long as my strength lasts, in a more or less sideways position."

"Then," said the Keel-Surface, "I get a pressure of air all on one side, and as there is, in effect, most of me towards the tail, the latter gets pressed sideways, and the Aeroplane thus tends to assume its first position and course."

"I see," said Efficiency, and, daintily holding the Chalk, she approached the Blackboard. "Is this what you mean?"

"Yes, that's right enough," said the Keel-Surface, "and you might remember, too, that I always make the Aeroplane nose into the gusts rather than away from them."

"If that was not the case," broke in Lateral Stability, and affecting the fashionable Flying Corps stammer, "it would be a h-h-h-o-r-rible affair! If there were too much Keel-Surface in front, then that gust would blow the Aeroplane round the other way a very considerable distance. And the right-hand Surface being on the outside of the turn would have more speed, and consequently more Lift, than the Surface on the other side. That means a greater proportion of the Lift on that side, and before you could say Warp to the Ailerons over the Aeroplane would go—probable result a bad side-slip" (see illustration A, over-leaf).

"And what can the Pilot do to save such a situation as that?" said Efficiency.

"Well," replied Lateral Stability, "he will try to turn the Aeroplane sideways and back to an even keel by means of warping the Ailerons or little wings which are hinged on to the Wing-tips, and about which you will hear more later on; but if the side-slip is very bad he may not be able to right the Aeroplane by means of the Ailerons, and then the only thing for him to do is to use the Rudder and to turn the nose of the Aeroplane down and head-on to the direction of motion. The Aeroplane will then be meeting the air in the direction it is designed to do so, and the Surfaces and also the controls (the Rudder, Ailerons, and Elevator) will be working efficiently; but its attitude relative to the earth will probably be more or less upside-down, for the action of turning the Aeroplane's nose down results, as you will see by the illustration B, in the right wing, which is on the

THE AEROPLANE SPEAKS

outside of the circle. travelling through the air with greater speed than the left-hand wing. More Speed means more Lift, so that results in overturning the Aeroplane still more; but now it is, at any rate, meeting the air as it is designed

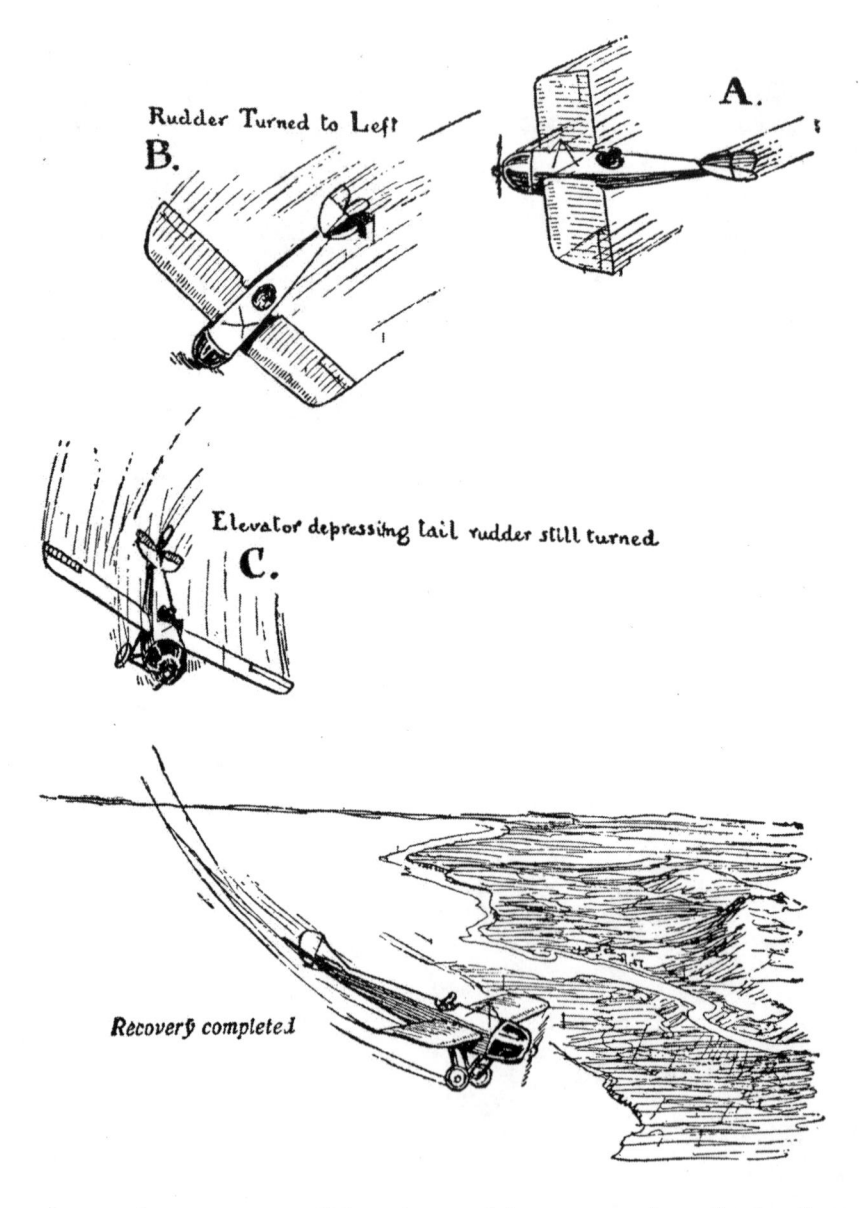

Rudder Turned to Left
B.

A.

Elevator depressing tail rudder still turned
C.

Recovery completed

to meet it, and everything is working properly. It is then only necessary to warp the Elevator, as shown in illustration C, in order to bring the Aeroplane into a proper attitude relative to the earth."

THE PRINCIPLES FINISH THE JOB

"Ah!" said the Rudder, looking wise, "it's in a case like that when I become the Elevator and the Elevator becomes me."

"That's absurd nonsense," said the Blackboard, "due to looseness of thought and expression."

"Well," replied the Rudder, "when the Aeroplane is in position A and I am used, then I depress or *elevate* the nose of the machine; and, if the Elevator is used, then it turns the Aeroplane to right or left, which is normally my function. Surely our *rôles* have changed one with the other, and I'm then the Elevator and the Elevator is me!"

Said Lateral Stability to the Rudder, "That's altogether the wrong way of looking at it, though I admit"—and

this rather sarcastically—"that the way you put it sounds rather fine when you are talking of your experiences in the air to those 'interested in aviation' but knowing little about it; but it won't go down here! You are a Controlling Surface designed to turn the Aeroplane about its vertical axis, and the Elevator is a Controlling Surface designed to turn the Aeroplane about its lateral axis. Those are your respective jobs, and you can't possibly change them about. Such talk only leads to confusion, and I hope we shall hear no more of it."

"Thanks," said Efficiency to Lateral Stability. "And now, please, will you explain your duties?"

THE AEROPLANE SPEAKS

"My duty is to keep the Aeroplane horizontal from Wing-tip to Wing-tip. First of all, I sometimes arrange with the Rigger to *wash-out,* that is decrease, the Angle of Incidence on one side of the Aeroplane, and to effect the reverse condition, if it is not too much trouble, on the other side."

"But," objected Efficiency, "the Lift varies with the Angle of Incidence, and surely such a condition will result in one side of the Aeroplane lifting more than the other side?"

"That's all right," said the Propeller, "it's meant to off-set the tendency of the Aeroplane to turn over sideways in the opposite direction to which I revolve."

"That's quite clear, though rather unexpected; but how do you counteract the effect of the gusts when they try to overturn the Aeroplane sideways?" said she, turning to Lateral Stability again.

"Well," he replied, rather miserably, "I'm not nearly so perfect as the Longitudinal and Directional Stabilities. The Dihedral Angle—that is, the upward inclination of the Surfaces towards their wing-tips—does what it can for me, but, in my opinion, it's a more or less futile effort. The Blackboard will show you the argument." And he at once showed them two Surfaces, each set at a Dihedral Angle like this:

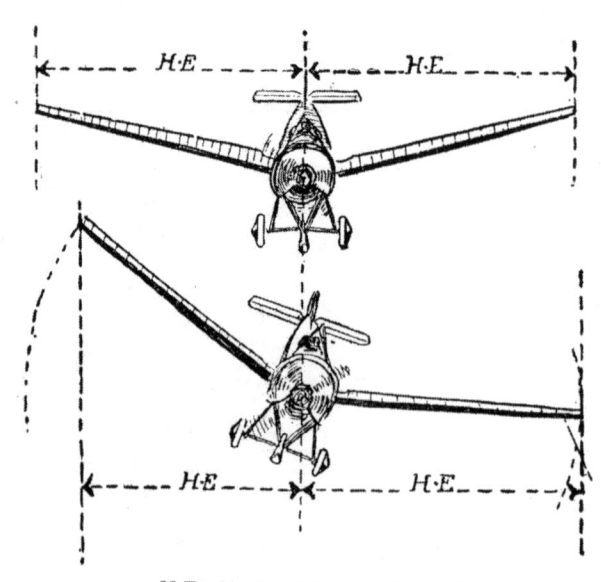

H.E., Horizontal equivalent.

THE PRINCIPLES FINISH THE JOB

"Please imagine," said the Blackboard, "that the top V is the front view of a Surface flying towards you. Now if a gust blows it into the position of the lower V you see that the horizontal equivalent of the Surface on one side becomes larger, and on the other side it becomes smaller. That results in more Lift on the lower side and less on the higher side, and if the V is large enough it should produce such a difference in the Lift of one side to the other as to quickly turn the Aeroplane back to its former and normal position."

"Yes," said the Dihedral Angle, "that's what would happen if they would only make me large enough; but they won't do it because it would too greatly decrease the horizontal equivalent, and therefore the Lift, and incidentally it would, as Aeroplanes are built to-day, produce an excess of Keel Surface above the turning axis, and that in itself would spoil the Lateral Stability. The Keel Surface should be equally divided above and below the longitudinal turning axis (upon which the Aeroplane rolls sideways), or the side upon which there is an excess will get blown over by the gusts. It strikes me that my future isn't very promising, and about my only chance is when the Junior Draughtsman makes a mistake, as he did the other day. And just think of it, they call him a Designer now that he's got a job at the Factory! What did he do? Why, he calculated the weights wrong and got the Centre of Gravity too high, and they didn't discover it until the machine was built. Then all they could do was to give me a larger Angle. That dropped the bottom of the V lower down, and as that's the centre of the machine, where all the Weight is, of course that put the Centre of Gravity in its right place. But now there is too much Keel Surface above, and the whole thing's a Bad Compromise, not at all like Our Efficiency."

And Efficiency, blushing very prettily at the compliment, then asked, "And how does the Centre of Gravity affect matters?"

"That's easy," said Grandfather Gravity. "I'm so heavy that if I am too low down I act like a pendulum and cause the Aeroplane to roll about sideways, and if I am too high I'm like a stick balanced on your finger, and then if I'm disturbed, over I go and the Aeroplane with me; and, in addition to that, there are the tricks I play with the Aeroplane when it's banked up,* *i.e.*, tilted sideways for a turn, and Centrifugal Force sets me going the way I'm not wanted to go. No; I get on best with Lateral Stability when my Centre is right on the centre of Drift, or, at any rate, not much below it." And with that he settled back into the Lecturer's Chair and went sound asleep again, for he was so very, very old, in fact the father of all the Principles.

And the Blackboard had been busy, and now showed them a picture of the Aeroplane as far as they knew it, and

you will see that there is a slight Dihedral Angle, and also, fixed to the tail, a vertical Keel Surface or *fin*, as is very often the case in order to ensure the greater effect of such surface being behind the vertical turning axis.

But Efficiency, growing rather critical with her newly gained knowledge, cried out: "But where's the horizontal Tail Surface? It doesn't look right like that!"

"This is when I have the pleasure of meeting you, my dear," said Longitudinal Stability. "Here's the Tail Surface," he said, "and in order to help me it must be set *in effect* at a much less Angle of Incidence than the Main Surface.

---

* Banking: When an aeroplane is turned to the left or the right the centrifugal force of its momentum causes it to skid sideways and outwards away from the centre of the turn. To minimize such action the pilot banks, *i.e.*, tilts, the aeroplane sideways in order to oppose the underside of the planes to the air. The aeroplane will not then skid outwards beyond the slight skid necessary to secure a sufficient pressure of air to balance the centrifugal force.

THE PRINCIPLES FINISH THE JOB

To explain we must trouble the Blackboard again," and this was his effort:

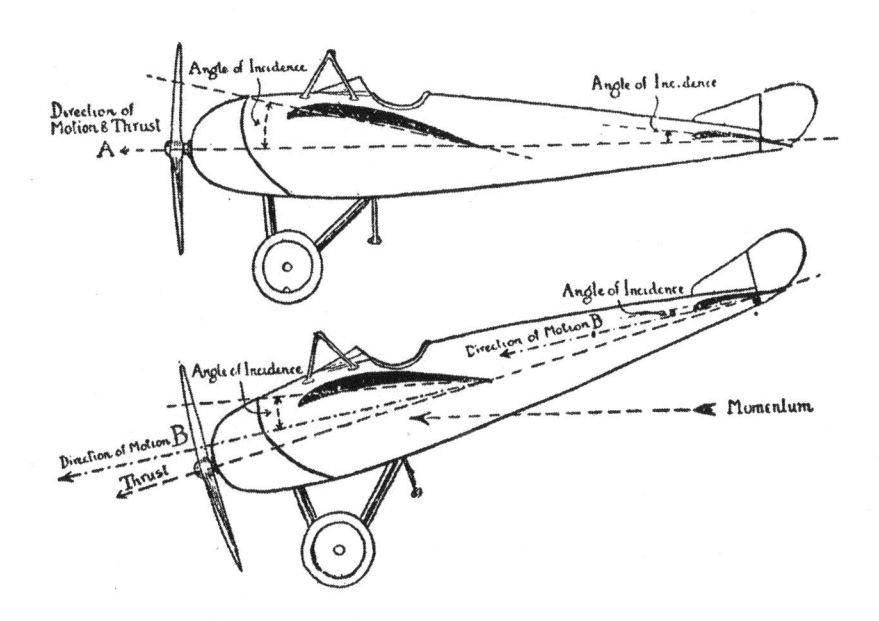

"I have tried to make that as clear as possible," he said. "It may appear a bit complicated at first, but if you will take the trouble to look at it for a minute you will find it quite simple. A is the normal and proper direction of motion of the Aeroplane, but, owing to a gust of air, it takes up the new nose-down position. Owing to Momentum, however, it does not fly straight along in that direction, but moves more or less in the direction B, which is the resultant of the two forces, Momentum and Thrust. And so you will note that the Angle of Incidence, which is the inclination of the Surfaces to the Direction of Motion, has decreased, and of course the Lift decreases with it. You will also see, and this is the point, that the Tail Surface has lost a higher proportion of its Angle, and consequently its Lift, than has the Main Surface. Then, such being the case, the Tail must fall and the Aeroplane assume its normal position again, though probably at a slightly lower altitude."

"I'm afraid I'm very stupid," said Efficiency, "but please tell me why you lay stress upon the words '*in effect.*'"

THE AEROPLANE SPEAKS

"Ah! I was wondering if you would spot that," he replied. "And there is a very good reason for it. You see, in some Aeroplanes the Tail Surface may be actually set at the same Angle on the machine as the Main Surface, but owing to the air being deflected downwards by the front Main Surface it meets the Tail Surface at a lesser angle, and indeed in some cases at no angle at all. The Tail is then for its surface getting less Lift than the Main Surface, although set at the same angle on the machine. It may then be said to have *in effect* a less Angle of Incidence. I'll just show you on the Blackboard."

"And now," said Efficiency, "I have only to meet the Ailerons and the Rudder, haven't I?"

"Here we are," replied the Ailerons, or little wings. "Please hinge us on to the back of the Main Surfaces, one of us at each Wing-tip, and join us up to the Pilot's joystick by means of the control cables. When the Pilot wishes to tilt the Aeroplane sideways, he will move the stick and depress us upon one side, thus giving us a larger Angle of Incidence and so creating more Lift on that side of the Aeroplane; and, by means of a cable connecting us with the Ailerons on the other side of the Aeroplane, we shall, as we are depressed, pull them up and give them a reverse or negative Angle of

THE PRINCIPLES FINISH THE JOB

Incidence, and that side will then get a reverse Lift or downward thrust, and so we are able to tilt the Aeroplane sideways.

"And we work best when the Angle of Incidence of the Surface in front of us is very small, for which reason it is sometimes decreased or *washed-out* towards the Wing-tips. The reason of that is that by the time the air reaches us it has been deflected downwards—the greater the Angle of Incidence the more it is driven downwards—and in order for us to secure a Reaction from it, we have to take such a

"Wash out" on both sides.

large Angle of Incidence that we produce a poor proportion of Lift to Drift; but the smaller the Angle of the Surface in front of us the less the air is deflected downwards, and consequently the less Angle is required of us, and the better our proportion of Lift to Drift, which, of course, makes us much more effective Controls."

"Yes," said the Lateral and Directional Stabilities in one voice, "that's so, and the wash-out helps us also, for then the Surfaces towards their Wing-tips have less Drift or 'Head-Resistance,' and consequently the gusts will affect them and us less; but such decreased Angle of Incidence means decreased Lift as well as Drift, and the Designer does not always care to pay the price."

"Well," said the Ailerons, "if it's not done it will mean more work for the Rudder, and that won't please the Pilot."

"Whatever do you mean?" asked Efficiency. "What can the Rudder have to do with you?"

"It's like this," they replied: "when we are deflected downwards we gain a larger Angle of Incidence and also enter an area of compressed air, and so produce more Drift than those of us on the other side of the Aeroplane, which are deflected upwards into an area of rarefied air due to the *suction* effect (though that term is not academically correct) on the top of the Surface. If there is more Drift, *i.e.*, Resistance, on one side of the Aeroplane than on the other side, then of course it will turn off its course, and if that difference in Drift is serious, as it will very likely be if there is no wash-out, then it will mean a good deal of work for the Rudder in keeping the Aeroplane on its course, besides creating extra Drift in doing so."

"I think, then," said Efficiency, "I should prefer to have that wash-out,* and my friend the Designer is so clever at producing strength of construction for light weight, I'm pretty sure he won't mind paying the price in Lift. And now let me see if I can sketch the completed Aeroplane."

"Well, I hope that's all as it should be," she concluded, "for to-morrow the Great Test in the air is due."

* An explanation of the way in which the wash-out is combined with a wash-in to offset propellor torque will be found on p. 82.

PART III

THE GREAT TEST

It is five o'clock of a fine calm morning, when the Aeroplane is wheeled out of its shed on to the greensward of the Military Aerodrome. There is every promise of a good flying day, and, although the sun has not yet risen, it is light enough to discern the motionless layer of fleecy clouds some five thousand feet high, and far, far above that a few filmy mottled streaks of vapour. Just the kind of morning beloved of pilots.

A brand new, rakish, up-to-date machine it is, of highly polished, beautifully finished wood, fabric as tight as a drum, polished metal, and every part so perfectly "streamlined" to minimize Drift, which is the resistance of the air to the passage of the machine, that to the veriest tyro the remark of the Pilot is obviously justified.

"Clean looking 'bus, looks almost alive and impatient to be off. Ought to have a turn for speed with those lines."

"Yes," replies the Flight-Commander, "it's the latest of its type and looks a beauty. Give it a good test. A special report is required on this machine."

The A.M.'s* have now placed the Aeroplane in position facing the gentle air that is just beginning to make itself evident; the engine Fitter, having made sure of a sufficiency of oil and petrol in the tanks, is standing by the Propeller; the Rigger, satisfied with a job well done, is critically "vetting" the machine by eye, four A.M.'s are at their posts, ready to hold the Aeroplane from jumping the blocks which have been placed in front of the wheels; and the Flight-Sergeant is awaiting the Pilot's orders.

As the Pilot approaches the Aeroplane the Rigger springs

* A.M.'s: Air Mechanics.

to attention and reports, "All correct, sir," but the Fitter does not this morning report the condition of the Engine, for well he knows that this Pilot always personally looks after the preliminary engine test. The latter, in leathern kit, warm flying boots and goggled, climbs into his seat, and now, even more than before, has the Aeroplane an almost living appearance, as if straining to be off and away. First he moves the Controls to see that everything is clear, for sometimes when the Aeroplane is on the ground the control lever or "joy-stick" is lashed fast to prevent the wind from blowing the controlling surfaces about and possibly damaging them.

The air of this early dawn is distinctly chilly, and the A.M.'s are beginning to stamp their cold feet upon the dewy grass, but very careful and circumspect is the Pilot, as he mutters to himself, "Don't worry and flurry, or you'll die in a hurry."

At last he fumbles for his safety belt, but with a start remembers the Pilot Air Speed Indicator, and, adjusting it to zero, smiles as he hears the Pilot-head's gruff voice, "Well, I should think so, twenty miles an hour I was registering. That's likely to cause a green pilot to stall the Aeroplane. Pancake, they call it." And the Pilot, who is an old hand and has learned a lot of things in the air that mere earth-dwellers know nothing about, distinctly heard the Pilot Tube, whose mouth is open to the air to receive its pressure, stammer. "Oh Lor! I've got an earwig already—hope to goodness the Rigger blows me out when I come down—and this morning air simply fills me with moisture; I'll never keep the Liquid steady in the Gauge. I'm not sure of my rubber connections either."

"Oh, shut up!" cry all the Wires in unison, "haven't we got our troubles too? We're in the most horrible state of tension. It's simply murdering our Factor of Safety, and how we can possibly stand it when we get the Lift only the Designer knows."

"That's all right," squeak all the little Wire loops, "we're that accommodating, we're sure to elongate a bit and so relieve your tension." For the whole Aeroplane is braced together with innumerable wires, many of which

THE GREAT TEST

are at their ends bent over in the form of loops in order to connect with the metal fittings on the spars and elsewhere—a cheap and easy way of making connection.

"Elongate, you little devils, would you?" fairly shout the Angles of Incidence, Dihedral and Stagger, amid a chorus of groans from all parts of the Aeroplane. "What's going to happen to us then? How are we going to keep our adjustments upon which good flying depends?"

"Butt us and screw us,"* wail the Wires. "Butt us and screw us, and death to the Loops. That's what we sang to the Designer, but he only looked sad and scowled at the Directors."

"And who on earth are they?" asked the Loops, trembling for their troublesome little lives.

"On earth indeed," sniffed Efficiency, who had not spoken before, having been rendered rather shy by being badly compromised in the Drawing Office. "I'd like to get some of them up between Heaven and Earth, I would. I'd give 'em something to think of besides their Debits and Credits—but all the same the Designer will get his way in the end. I'm his Best Girl, you know, and if we could only get rid of the Directors, the little Tin god, and the Man-who-takes-the-credit, we should be quite happy." Then she abruptly subsides, feeling that perhaps the less said the better until she has made a reputation in the Air. The matter of that Compromise still rankled, and indeed it does seem hardly fit that a bold bad Tin god should flirt with Efficiency. You see there was a little Tin god, and he said "Boom, Boom Boom! Nonsense! It MUST be done," and things like that in a very loud voice, and the Designer tore his hair and was furious, but the Directors, who were thinking of nothing but Orders and Dividends, had the whip-hand of *him*, and so there you are, and so poor beautiful Miss Efficiency was compromised.

All this time the Pilot is carefully buckling his belt and making himself perfectly easy and comfortable, as all good pilots do. As he straightens himself up from a careful

---

* Butt means to thicken at the end. Screw means to machine a thread on the butt-end of the wire, and in this way the wire can make connection with the desired place by being screwed into a metal fitting, thus eliminating the disadvantage of the unsatisfactory loop.

THE AEROPLANE SPEAKS

inspection of the Deviation Curve* of the Compass and takes command of the Controls, the Throttle and the Ignition, the voices grow fainter and fainter until there is nothing but a trembling of the Lift and Drift wires to indicate to his understanding eye their state of tension in expectancy of the Great Test.

"Petrol on?" shouts the Fitter to the Pilot.

"Petrol on," replies the Pilot.

"Ignition off?"

"Ignition off."

Round goes the Propeller, the Engine sucking in the Petrol Vapour with satisfied gulps. And then—

"Contact?" from the Fitter.

"Contact," says the Pilot.

Now one swing of the Propeller by the Fitter, and the Engine is awake and working. Slowly at first though, and in a weak voice demanding, "Not too much Throttle, please. I'm very cold and mustn't run fast until my Oil has thinned and is circulating freely. Three minutes slowly, as you love me, Pilot."

Faster and faster turn the Engine and Propeller, and the Aeroplane, trembling in all its parts, strains to jump the blocks and be off. Carefully the Pilot listens to what the Engine Revolution Indicator says. At last, "Steady at 1,500 revs. and I'll pick up the rest in the Air." Then does he throttle down the Engine, carefully putting the lever back to the last notch to make sure that in such position the Throttle is still sufficiently open for the Engine to continue working, as otherwise it might lead to him "losing" his Engine in the air when throttling down the power for descent. Then, giving the official signal, he sees the blocks removed from the wheels, and the Flight-Sergeant saluting he knows that all is clear to ascend. One more signal, and all the A.M.'s run clear of the Aeroplane.

Then gently, gently mind you, with none of the "crashing on" bad Pilots think so fine, he opens the Throttle and, the Propeller Thrust overcoming its enemy the Drift, the Aeroplane moves forward.

"Ah!" says the Wind-screen, "that's Discipline, that

---

*Deviation curve: A curved line indicating any errors in the compass.

THE GREAT TEST

is. Through my little window I see most things, and don't I just know that poor discipline always results in poor work in the air, and don't you forget it."

"Discipline is it?" complains the Under-carriage, as its wheels roll swiftly over the rather rough ground. "I'm *bump* getting it, and *bump, bump,* all I want, *bang, bump, rattle,* too!" But, as the Lift increases with the Speed, the complaints of the Under-carriage are stilled, and then, the friendly Lift becoming greater than the Weight, the Aeroplane swiftly and easily takes to the air.

Below is left the Earth with all its bumps and troubles. Up into the clean clear Air moves with incredible speed and steadiness this triumph of the Designer, the result of how much mental effort, imagination, trials and errors, failures and successes, and many a life lost in high endeavour.

Now is the mighty voice of the Engine heard as he turns the Propeller nine hundred times a minute. Now does the Thrust fight the Drift for all it's worth, and the Air Speed Indicator gasps with delight, "One hundred miles an hour!"

And now does the burden of work fall upon the Lift and Drift Wires, and they scream to the Turnbuckles whose business it is to hold them in tension, "This is the limit! the Limit! THE LIMIT! Release us, if only a quarter turn." But the Turnbuckles are locked too fast to turn their eyes or utter a word. Only the Locking Wires thus: "Ha! ha! the Rigger knew his job. He knew the trick, and there's no release here." For an expert rigger will always use the locking wire in such a way as to oppose the slightest tendency of the turnbuckle to unscrew. The other kind of rigger will often use the wire in such a way as to allow the turnbuckle, to the "eyes" of which the wires are attached, to unscrew a quarter of a turn or more, with the result that the correct adjustment of the wires may be lost; and upon their fine adjustment much depends.

And the Struts and the Spars groan in compression and pray to keep straight, for once "out of truth" there is, in addition to possible collapse, the certainty that in bending they will throw many wires out of adjustment.

And the Fabric's quite mixed in its mind, and ejaculates,

"Now, who would have thought I got more Lift from the top of the Surface than its bottom?" And then truculently to the Distance Pieces, which run from rib to rib, "Just keep the Ribs from rolling, will you? or you'll see me strip. I'm an Irishman, I am, and if my coat comes off—— Yes, Irish, I said. I used to come from Egypt, but I've got naturalized since the War began."

Then the Air Speed Indicator catches the eye of the Pilot. "Good enough," he says as he gently deflects the Elevator and points the nose of the Aeroplane upwards in search of the elusive Best Climbing Angle.

"Ha! ha!" shouts the Drift, growing stronger with the increased Angle of Incidence. "Ha! ha!" he laughs to the Thrust. "Now I've got you. Now who's Master?"

And the Propeller shrieks hysterically, "Oh! look at me. I'm a helicopter. That's not fair. Where's Efficiency?" And she can only sadly reply, "Yes, indeed, but you see we're a Compromise."

And the Drift has hopes of reaching the Maximum Angle of Incidence and vanquishing the Thrust and the Lift. And he grows very bold as he strangles the Thrust; but the situation is saved by the Propeller, who is now bravely helicopting skywards, somewhat to the chagrin of Efficiency.

"Much ado about nothing," quotes the Aeroplane learnedly. "Compromise or not, I'm climbing a thousand feet a minute. Ask the Altimeter. He'll confirm it."

And so indeed it was. The vacuum box of the Altimeter was steadily expanding under the decreased pressure of the rarefied air, and by means of its little levers and its wonderful chain no larger than a hair it was moving the needle round the gauge and indicating the ascent at the rate of a thousand feet a minute.

And lo! the Aeroplane has almost reached the clouds! But what's this? A sudden gust, and down sinks one wing and up goes the other. "Oh, my Horizontal Equivalent!" despairingly call the Planes; "it's eloping with the Lift, and what in the name of Gravity will happen? Surely there was enough scandal in the Factory without this, too!" For the lift varies with the horizontal equivalent of the planes, so that if the aeroplane tilts sideways beyond a certain angle, the lift becomes less than the weight of the machine, which must then fall. A fall in such a position is known as a "side-slip."

But the ever-watchful Pilot instantly depresses one aileron, elevating the other, with just a touch of the rudder to keep on the course, and the Planes welcome back their precious Lift as the Aeroplane flicks back to its normal position.

"Bit bumpy here under these clouds," is all the Pilot says as he heads for a gap between them, and the next minute the Aeroplane shoots up into a new world of space.

"My eye!" ejaculates the Wind-screen, "talk about a view!" And indeed mere words will always fail to express the wonder of it. Six thousand feet up now, and look! The sun is rising quicker than ever mortal on earth witnessed its ascent. Far below is Mother Earth, wrapt in mists and deep blue shadows, and far above are those light, filmy, ethereal clouds now faintly tinged with pink. And all about great mountains of cloud, lazily floating in space. The sun rises and they take on all colours, blending one with the other, from dazzling white to crimson and deep violet-blue. Lakes and rivers here and there in the enormous expanse of country below refract the level rays of the sun and, like so many immense diamonds, send dazzling shafts of light far upwards. The tops of the hills now laugh to the light of the sun, but the valleys are still mysterious dark blue caverns, crowned with white filmy lace-like streaks of vapour. And withal the increasing sense with altitude of vast, clean, silent solitudes of space.

Lives there the man who can adequately describe this Wonder? "Never," says the Pilot, who has seen it many times, but to whom it is ever new and more wonderful.

Up, up, up, and still up, unfalteringly speeds the Pilot and his mount. Sweet the drone of the Engine and steady the Thrust as the Propeller exultingly battles with the Drift.

And look! What is that bright silver streak all along the horizon? It puzzled the Pilot when first he saw it, but now he knows it for the Sea, full fifty miles away!

And on his right is the brightness of the Morn and the smiling Earth unveiling itself to the ardent rays of the Sun; and on his left, so high is he, there is yet black Night, hiding

THE AEROPLANE SPEAKS

innumerable Cities, Towns, Villages and all those places where soon teeming multitudes of men shall awake, and by their unceasing toil and the spirit within them produce marvels of which the Aeroplane is but the harbinger.

And the Pilot's soul is refreshed, and his vision, now exalted, sees the Earth a very garden, even as it appears at that height, with discord banished and a happy time come, when the Designer shall have at last captured Efficiency, and the Man-who-takes-the-credit is he who has earned it, and when kisses are the only things that go by favour.

Now the Pilot anxiously scans the Barograph, which is an instrument much the same as the Altimeter; but in this case the expansion of the vacuum box causes a pen to trace a line upon a roll of paper. This paper is made by clockwork to pass over the point of the pen, and so a curved line is made which accurately registers the speed of the ascent in feet per minute. No longer is the ascent at the rate of a thousand feet a minute, and the Propeller complains to the Engine, "I'm losing my Revs. and the Thrust. Buck up with the Power, for the Lift is decreasing, though the Weight remains much the same."

Quoth the Engine: "I strangle for Air. A certain proportion, and that of right density, I must have to one part of Petrol, in order to give me full power and compression, and here at an altitude of ten thousand feet the Air is only two-thirds as dense as at sea-level. Oh, where is he who will invent a contrivance to keep me supplied with Air of right density and quality? It should not be impossible within certain limits."

"We fully agree," said the dying Power and Thrust. "Only maintain Us and you shall be surprised at the result. For our enemy Drift *decreases in respect of distance with the increase of altitude and rarity of air,* and there is no limit to the speed through space if only our strength remains. And with oxygen for Pilot and Passengers and a steeper pitch*

---

* A propeller screws through the air, and the distance it advances during one revolution, supposing the air to be solid, is known as the pitch. The pitch, which depends upon the angle of the propeller blades, must be equal to the speed of the aeroplane, plus the slip, and if, on account of the rarity of the air, the speed of the aeroplane increases, then the angle and pitch should be correspondingly increased. Propellers with a pitch capable of being varied by the pilot are the dream of propeller designers. For explanation of "slip" see Chapter IV. on propellers.

THE GREAT TEST

for the Propeller we may then circle the Earth in a day!"

Ah, Reader, smile not unbelievingly, as you smiled but a few years past. There may be greater wonders yet. Consider that as the speed increases, so does the momentum or stored-up force in the mass of the aeroplane become terrific. And, bearing that in mind, remember that with altitude *gravity decreases*. There may yet be literally other worlds to conquer.*

Now at fifteen thousand feet the conditions are chilly and rare, and the Pilot, with thoughts of breakfast far below, exclaims, "High enough! I had better get on with the Test." And then, as he depresses the Elevator, the Aeroplane with relief assumes its normal horizontal position. Then, almost closing the Throttle, the Thrust dies away. Now, the nose of the Aeroplane should sink of its own volition, and the craft glide downward at flying speed, which is in this case a hundred miles an hour. That is what should happen if the Designer has carefully calculated the weight of every part and arranged for the centre of gravity to be just the right distance in front of the centre of lift. Thus is the Aeroplane "nose-heavy" as a glider, and just so to a degree ensuring a speed of glide equal to its flying speed. And the Air Speed Indicator is steady at one hundred miles an hour, and "That's all right!" exclaims the Pilot. "And very useful, too, in a fog or a cloud," he reflects, for then he can safely leave the angle of the glide to itself, and give all his attention, and he will need it all, to keeping the Aeroplane horizontal from wing-tip to wing-tip, and to keeping it straight on its course. The latter he will manage with the rudder, controlled by his feet, and the Compass will tell him whether a straight course is kept. The former he will control by the Ailerons, or little wings hinged to the tips of the planes, and the bubble in the Inclinometer in front of him must be kept in the middle.

A Pilot, being only human, may be able to do two things at once, but three is a tall order, so was this Pilot relieved

---

*Getting out of my depth? Invading the realms of fancy? Well, perhaps so, but at any rate it is possible that extraordinary speed through space may be secured if means are found to maintain the impulse of the engine and the thrust-drift efficiency of the propeller at great altitude.

THE AEROPLANE SPEAKS

to find the Design not at fault and his craft a "natural glider." To correct this nose-heavy tendency when the Engine is running, and descent not required, the centre of Thrust is arranged to be a little below the centre of Drift or Resistance, and thus acts as a counter-balance.

But what is this stream of bad language from the Exhaust Pipe, accompanied by gouts of smoke and vapour? The Engine, now revolving at no more than one-tenth its normal speed, has upset the proportion of petrol to air, and combustion is taking place intermittently or in the Exhaust Pipe, where it has no business to be.

"Crash, Bang, Rattle——!——!——!" and worse than that, yells the Exhaust, and the Aeroplane, who is a gentleman and not a box kite,* remonstrates with the severity of a Senior Officer. "See the Medical Officer, you young Hun. Go and see a doctor. Vocal diarrhœa, that's your complaint, and a very nasty one too. Bad form, bad for discipline, and a nuisance in the Mess. What's your Regiment? Special Reserve, you say? Humph! Sounds like Secondhand Bicycle Trade to me!"

Now the Pilot decides to change the straight gliding descent to a spiral one, and, obedient to the Rudder, the Aeroplane turns to the left. But the Momentum (two tons at 100 miles per hour is no small affair) heavily resents this change of direction, and tries its level best to prevent it and to pull the machine sideways and outwards from its spiral course—that is, to make it "side-skid" outwards. But the Pilot deflects the Ailerons and "banks" up the planes to the correct angle, and, the Aeroplane skidding sideways and outwards, the lowest surfaces of the planes press up against the air until the pressure equals the centrifugal force of the Momentum, and the Aeroplane spirals steadily downwards.

Down, down, down, and the air grows denser, and the Pilot gulps largely, filling his lungs with the heavier air to counteract the increasing pressure from without. Down through a gap in the clouds, and the Aerodrome springs into view, appearing no larger than a saucer, and the Pilot, having by now got the "feel" of the Controls, proceeds

* Box-kite. The first crude form of biplane.

THE GREAT TEST

to put the Aeroplane through its paces. First at its Maximum Angle, staggering along tail-down and just maintaining horizontal flight; then a dive at far over flying speed, finishing with a perfect loop; then sharp turns with attendant vertical "banks" and then a wonderful switchback flight, speeding down at a hundred and fifty miles an hour with short, exhilarating ascents at the rate of two thousand feet a minute!

All the parts are now working well together. Such wires as were before in undue tension have secured relief by slightly elongating their loops, and each one is now doing its bit, and all are sharing the burden of work together.

The Struts and the Spars, which felt so awkward at first, have bedded themselves in their sockets, and are taking the compression stresses uncomplainingly.

The Control Cables of twisted wire, a bit tight before, have slightly lengthened by perhaps the eighth of an inch, and, the Controls instantly responding to the delicate touch of the Pilot, the Aeroplane, at the will of its Master, darts this way and that way, dives, loops, spirals, and at last, in one long, magnificent glide, lands gently in front of its shed.

"Well, what result?" calls the Flight-Commander to the Pilot.

"A hundred miles an hour and a thousand feet a minute," he briefly replies.

"And a very good result too," says the Aeroplane, complacently, as he is carefully wheeled into his shed.

* * * * * *

That is the way Aeroplanes speak to those who love them and understand them. Lots of Pilots know all about it, and can spin you wonderful yarns, much better than this one, if you catch them in a confidential mood—on leave, for instance, and after a good dinner.

PART IV

'CROSS COUNTRY

The Aeroplane had been designed and built, and tested in the air, and now stood on the Aerodrome ready for its first 'cross-country flight.

It had run the gauntlet of pseudo-designers, crank inventors, press "experts," and politicians; of manufacturers keen on cheap work and large profits; of poor pilots who had funked it, and good pilots who had expected too much of it. Thousands of pounds had been wasted on it, many had gone bankrupt over it, and others it had provided with safe fat jobs.

Somehow, and despite every conceivable obstacle, it had managed to muddle through, and now it was ready for its work. It was not perfect, for there were fifty different ways in which it might be improved, some of them shamefully obvious. But it was fairly sound mechanically, had a little inherent stability, was easily controlled, could climb a thousand feet a minute, and its speed was a hundred miles an hour. In short, quite a creditable machine, though of course the right man had not got the credit.

It is rough, unsettled weather with a thirty mile an hour wind on the ground, and that means fifty more or less aloft. Lots of clouds at different altitudes to bother the Pilot, and the air none to clear for the observation of landmarks.

As the Pilot and Observer approach the Aeroplane the former is clearly not in the best of tempers. "It's rotten luck," he is saying, "a blank shame that I should have to take this blessed 'bus and join X Reserve Squadron, stationed a hundred and fifty miles from anywhere; and just as I have licked my Flight into shape. Now some slack blighter will, I suppose, command it and get the credit of all my work!"

'CROSS COUNTRY

"Shut up, you grouser," said the Observer. "Do you think you're the only one with troubles? Haven't I been through it too? Oh! I know all about it! You're from the Special Reserve and your C.O. doesn't like your style of beauty, and you won't lick his boots, and you were a bit of a technical knut in civil life, but now you've jolly well got to know less than those senior to you. Well! It's a jolly good experience for most of us. Perhaps conceit won't be at quite such a premium after this war. And what's the use of grousing? That never helped anyone. So buck up, old chap. Your day will come yet. Here's our machine, and I must say it looks a beauty!"

And, as the Pilot approaches the Aeroplane, his face brightens and he soon forgets his troubles as he critically inspects the craft which is to transport him and the Observer over the hills and far away. Turning to the Flight-Sergeant he inquires, "Tank full of petrol and oil?"

"Yes, sir," he replies, "and everything else all correct. Propeller, engine, and body covers on board, sir; tool kit checked over and in the locker; engine and Aeroplane logbooks written up, signed, and under your seat; engine revs. up to mark, and all the control cables in perfect condition and tension."

"Very good," said the Pilot; and then turning to the Observer, "Before we start you had better have a look at the course I have mapped out (see p. 40).

"A is where we stand and we have to reach B, a hundred and fifty miles due North. I judge that, at the altitude we shall fly, there will be an East wind, for although it is not quite East on the ground it is probably about twenty degrees different aloft, the wind usually moving round clockways to about that extent. I think that it is blowing at the rate of about fifty miles an hour, and I therefore take a line on the map to C, fifty miles due West of A. The Aeroplane's speed is a hundred miles an hour, and so I take a line of one hundred miles from C to D. Our compass course will then be in the direction A—E, which is always a line parallel to C—D. That is, to be exact, it will be fourteen degrees off the C—D course, as, in this part of the globe, there is that much difference between the North and South lines on the map and the magnetic North to which the compass needle points. If the compass has an error, as it may have of a

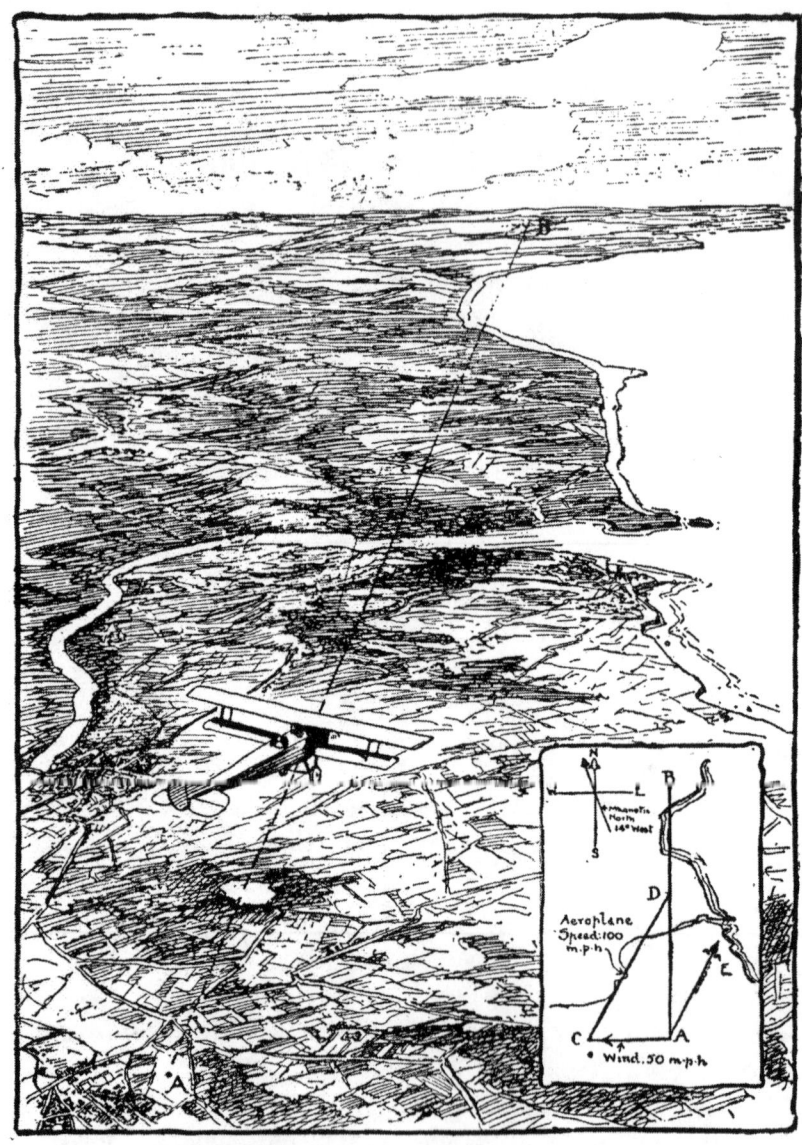

A—B, 150 miles.
A—C, 50 miles ; direction and miles per hour of wind.
C—D, 100 miles ; air speed of aeroplane.
A—D, Distance covered by aeroplane in one hour.
A—E, Compass course.

few degrees, that, too, must be taken into account, and the deviation or error curve on the dashboard will indicate it.

'CROSS COUNTRY

"The Aeroplane will then always be pointing in a direction parallel to A—E, but, owing to the side wind, it will be actually travelling over the course A—B, though in a rather sideways attitude to that course.

"The distance we shall travel over the A—B course in one hour is A—D. That is nearly eighty-seven miles, so we ought to accomplish our journey of a hundred and fifty miles in about one and three-quarter hours.

"I hope that's quite clear to you. It's a very simple way of calculating the compass course, and I always do it like that."

"Yes, that's plain enough. You have drafted what engineers call 'a parallelogram of forces'; but suppose you have miscalculated the velocity of the wind, or that it should change in velocity or direction?"

"Well, that of course will more or less alter matters," replies the Pilot. "But there are any number of good landmarks such as lakes, rivers, towns, and railway lines. They will help to keep us on the right course, and the compass will, at any rate, prevent us from going far astray when between them."

"Well, we'd better be off, old chap. Hop aboard." This from the Observer as he climbs into the front seat from which he will command a good view over the lower plane; and the Pilot takes his place in the rear seat, and, after making himself perfectly comfortable, fixing his safety belt, and moving the control levers to make sure that they are working freely, he gives the signal to the Engine Fitter to turn the propeller and so start the engine.

Round buzzes the Propeller, and the Pilot, giving the official signal, the Aeroplane is released and rolls swiftly over the ground in the teeth of the gusty wind.

In less than fifty yards it takes to the air and begins to climb rapidly upwards, but how different are the conditions to the calm morning of yesterday! If the air were visible it would be seen to be acting in the most extraordinary manner; crazily swirling, lifting and dropping, gusts viciously colliding—a mad phantasmagoria of forces!

Wickedly it seizes and shakes the Aeroplane; then tries to turn it over sideways; then instantly changes its mind and in a second drops it into a hole a hundred feet deep; and if it were not for his safety belt the Pilot might find his seat sinking away from beneath him.

Gusts strike the front of the craft like so many slaps in the face; and others, with the motion of mountainous waves, sometimes lift it hundreds of feet in a few seconds, hoping to see it plunge over the summit in a death-dive—and so it goes on, but the Pilot, perfectly at one with his mount and

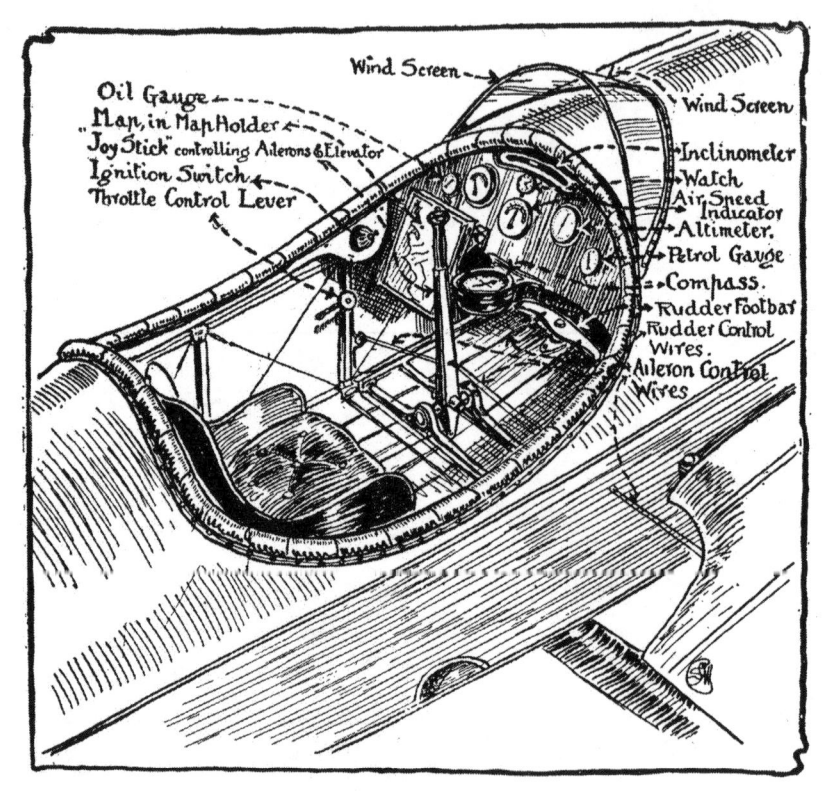

The Pilot's Cock-pit.

instantly alert to its slightest motion, is skilfully and naturally making perhaps fifty movements a minute of hand and feet; the former lightly grasping the "joy-stick" which controls the Elevator hinged to the tail, and also the Ailerons or little wings hinged to the wing-tips; and the latter moving the Rudder control-bar.

A strain on the Pilot? Not a bit of it, for this is his Work which he loves and excels in; and given a cool head, alert eye, and a sensitive touch for the controls, what sport can compare with these ever-changing battles of the air?

The Aeroplane has all this time been climbing in great wide circles, and is now some three thousand feet above the Aerodrome which from such height looks absurdly small. The buildings below now seem quite squat; the hills appear to have sunk away into the ground, and the whole country below, cut up into diminutive fields, has the appearance of having been lately tidied and thoroughly spring-cleaned! A doll's country it looks, with tiny horses and cows ornamenting the fields and little model motor-cars and carts stuck on the roads, the latter stretching away across the country like ribbons accidentally dropped.

At three thousand feet altitude the Pilot is satisfied that he is now sufficiently high to secure, in the event of engine failure, a long enough glide to earth to enable him to choose and reach a good landing-place; and, being furthermore content with the steady running of the engine, he decides to climb no more but to follow the course he has mapped out. Consulting the compass, he places the Aeroplane on the A—E course and, using the Elevator, he gives his craft its minimum angle of incidence at which it will just maintain horizontal flight and secure its maximum speed.

Swiftly he speeds away, and few thoughts he has now for the changing panorama of country, cloud, and colour. Ever present in his mind are the three great 'cross-country queries. "Am I on my right course? Can I see a good landing-ground within gliding distance?" And "How is the Engine running?"

Keenly both he and the Observer compare their maps with the country below. The roads, khaki-coloured ribbons, are easily seen but are not of much use, for there are so many of them and they all look alike from such an altitude.

Now where can that lake be which the map shows so plainly? He feels that surely he should see it by now, and has an uncomfortable feeling that he is flying too far West. What pilot is there indeed who has not many times experienced such unpleasant sensation? Few things in the air can create greater anxiety. Wisely, however, he sticks

THE AEROPLANE SPEAKS

to his compass course, and the next minute he is rewarded by the sight of the lake, though indeed he now sees that the direction of his travel will not take him over it, as should be the case if he were flying over the shortest route to his destination. He must have slightly miscalculated the velocity or direction of the side-wind.

"About ten degrees off," he mutters, and, using the Rudder, corrects his course accordingly.

Now he feels happier and that he is well on his way. The gusts, too, have ceased to trouble him as, at this altitude, they are not nearly so bad as they were near the ground, the broken surface of which does much to produce them; and sometimes for miles he makes but a movement or two of the controls.

The clouds just above race by with dizzy and uniform speed; the country below slowly unrolls, and the steady drone of the Engine is almost hypnotic in effect. "Sleep, sleep, sleep," it insidiously suggests. "Listen to me and watch the clouds; there's nothing else to do. Dream, dream, dream of speeding through space for ever, and ever, and ever; and rest, rest, rest to the sound of my rhythmical hum. Droning on and on, nothing whatever matters. All things now are merged into speed through space and a sleepy monotonous d-d-r-r-o-o-n-n-e - - - - -." But the Pilot pulls himself together with a start and peers far ahead in search of the next landmark. This time it is a little country town. red-roofed his map tells him, and roughly of cruciform shape; and, sure enough, there in the right direction are the broken outlines of a few red roofs peeping out from between the trees.

Another minute and he can see this little town, a fairy town it appears, nestling down between the hills with its red roofs and picturesque shape, a glowing and lovely contrast with the dark green of the surrounding moors.

So extraordinarily clean and tidy it looks from such a height, and laid out in such orderly fashion with perfectly defined squares, parks, avenues, and public buildings, it indeed appears hardly real, but rather as if it has this very day materialized from some delightful children's book!

Every city and town you must know has its distinct individuality to the Pilot's eye. Some are not fairy places

'CROSS COUNTRY

at all, but great dark ugly blots upon the fair countryside, and with tall shafts belching forth murky columns of smoke to defile clean space. Others, melancholy-looking masses of grey, slate-roofed houses, are always sad and dispirited; never welcoming the glad sunshine, but ever calling for leaden skies and a weeping Heaven. Others again, little coquettes with village green, white palings everywhere, bright gravel roads, and an irrepressible air of brightness and gaiety.

Then there are the rivers, silvery streaks peacefully winding far, far away to the distant horizon; they and the lakes the finest landmarks the Pilot can have. And the forests. How can I describe them? The trees cannot be seen separately, but merge altogether into enormous irregular dark green masses sprawling over the country, and sometimes with great ungainly arms half encircling some town or village; and the wind passing over the foliage at times gives the forest an almost living appearance, as of some great dragon of olden times rousing itself from slumber to devour the peaceful villages which its arms encircle.

And the Pilot and Observer fly on and on, seeing these things and many others which baffle my poor skill to describe—things, dear Reader, that you shall see, and poets sing of, and great artists paint in the days to come when the Designer has captured Efficiency. Then, and the time is near, shall you see this beautiful world as you have never seen it before, the garden it is, the peace it breathes, and the wonder of it.

The Pilot, flying on, is now anxiously looking for the railway line which midway on his journey should point the course. Ah! There it is at last, but suddenly (and the map at fault) it plunges into the earth! Well the writer remembers when that happened to him on a long 'cross-country flight in the early days of aviation. Anxiously he wondered "Are tunnels always straight?" and with what relief, keeping on a straight course, he picked up the line again some three miles farther on!

Now at last the Pilot sees the sea, just a streak on the north-eastern horizon, and he knows that his flight is two-thirds over. Indeed, he should have seen it before, but the air is none too clear, and he is not yet able to discern the river which soon should cross his path. As he swiftly speeds on the air becomes denser and denser with what he fears must be the beginning of a sea-fog, perhaps drifting inland along the course of the river. Now does he feel real anxiety, for it is the *duty* of a Pilot to fear fog, his deadliest enemy. Fog not only hides the landmarks by which he keeps his course, but makes the control of the Aeroplane a matter of the greatest difficulty. He may not realize it, but, in keeping his machine on an even keel, he is unconsciously balancing it against the horizon, and with the horizon gone he is lost indeed. Not only that, but it also prevents him from choosing his landing-place, and the chances are that, landing in a fog, he will smash into a tree, hedge, or building, with disastrous results. The best and boldest pilot 'wares a fog, and so this one, finding the conditions becoming worse and yet worse, and being forced to descend lower and lower in order to keep the earth within view, wisely decides to choose a landing-place while there is yet time to do so.

Throttling down the power of the engine he spirals downwards, keenly observing the country below. There are plenty of green fields to lure him, and his great object is to avoid one in which the grass is long, for that would bring his machine to a stop so suddenly as to turn it over; or one of rough surface likely to break the under-carriage. Now is perfect eyesight and a cool head indispensable. He sees and decides upon a field and, knowing his job, he sticks to that field with no change of mind to confuse him. It is none too large, and gliding just over the trees and head on to the wind he skilfully "stalls" his machine; that is, the speed having decreased sufficiently to avoid such a manœuvre resulting in ascent, he, by means of the Elevator, gives the Aeroplane as large an angle of incidence as possible, and the undersides of the planes meeting the air at such a large angle act as an air-brake, and the Aeroplane, skimming over the ground, lessens its speed and finally stops just at the farther end of the field.

Then, after driving the Aeroplane up to and under the lee of the hedge, he stops the engine, and quickly lashing the joy-stick fast in order to prevent the wind from blowing the controlling surfaces about and possibly damaging them, he hurriedly alights. Now running to the tail he lifts it up on to his shoulder, for the wind has become rough indeed and there is danger of the Aeroplane becoming unmanageable. By this action he decreases the angle at which the planes are inclined to the wind and so minimizes the latter's effect upon them. Then to the Observer, "Hurry up, old fellow, and try to find some rope, wire, or anything with which to picket the machine. The wind is rising and I shan't be able to hold the 'bus steady for long. Don't forget the wire-cutters. They're in the tool kit." And the Observer rushes off in frantic haste, before long triumphantly returning with a long length of wire from a neighbouring fence. Blocking up the tail with some débris at hand, they soon succeed, with the aid of the wire, in stoutly picketing the Aeroplane to the roots of the high hedge in front of it; done with much care, too, so that the wire shall not fray the fabric or set up dangerous bending-stresses in the woodwork. Their work is not done yet, for the Observer remarking, "I don't like the look of this thick weather and rather fear a heavy rainstorm," the Pilot replies, "Well, it's a fearful bore, but the first rule of our game is never to take an unnecessary risk, so out with the engine and body covers."

Working with a will they soon have the engine and the open part of the body which contains the seats, controls, and instruments snugly housed with their waterproof covers, and the Aeroplane is ready to weather the possible storm.

Says the Observer, "I'm remarkably peckish, and methinks I spy the towers of one of England's stately homes showing themselves just beyond that wood, less than a quarter of a mile away. What ho! for a raid. What do you say?"

"All right, you cut along and I'll stop here, for the Aeroplane must not be left alone. Get back as quickly as possible."

And the Observer trots off, leaving the Pilot filling his pipe and anxiously scrutinizing the weather conditions. Very thick it is now, but the day is yet young, and he has hopes of the fog lifting sufficiently to enable the flight to be resumed. A little impatiently he awaits the return of his comrade, but with never a doubt of the result, for the hospitality of the country house is proverbial among pilots! What old hand among them is there who cannot instance many a forced landing made pleasant by such hospitality? Never too late or too early to help with food, petrol, oil, tools, and assistants. Many a grateful thought has the writer for such kind help given in the days before the war (how long ago they seem!), when aeroplanes were still more imperfect than they are now, and involuntary descents often a part of 'cross-country flying.

Ah! those early days! How fresh and inspiring they were! As one started off on one's first 'cross-country flight, on a machine the first of its design, and with everything yet to learn, and the wonders of the air yet to explore; then the joy of accomplishment, the dreams of Efficiency, the hard work and long hours better than leisure; and what a field of endeavour—the realms of space to conquer! And the battle still goes on with ever-increasing success. Who is bold enough to say what its limits shall be?

So ruminates this Pilot-Designer, as he puffs at his pipe, until his reverie is abruptly disturbed by the return of the Observer.

"Wake up, you *airman*," the latter shouts. "Here's the very thing the doctor ordered! A basket of first-class grub and something to keep the fog out, too."

"Well, that's splendid, but don't call me newspaper names or you'll spoil my appetite!"

Then, with hunger such as only flying can produce, they appreciatively discuss their lunch, and with many a grateful thought for the donors—and they talk shop. They can't help it, and even golf is a poor second to flight talk. Says the Pilot, who must have his grievance, "Just observe where I managed to stop the machine. Not twenty feet from this hedge! A little more and we should have been through it and into Kingdom Come! I stalled as well as one could, but the tail touched the ground and so I could not give the Aeroplane any larger angle of incidence. Could I have given it a larger angle, then the planes would have become a much more effective air-brake, and we should have come to rest in a much shorter distance. It's all the fault of the tail. There's hardly a type of Aeroplane in existence in which the tail could not be raised several feet, and that would make all the difference. High tails mean a large angle of incidence when the machine touches ground and, with enough angle, I'll guarantee to safely land the fastest machine in a five-acre field. You can, I am sure, imagine what a difference that would make where forced landings are concerned!" Then rapidly sketching in his notebook, he shows the Observer the following illustration:

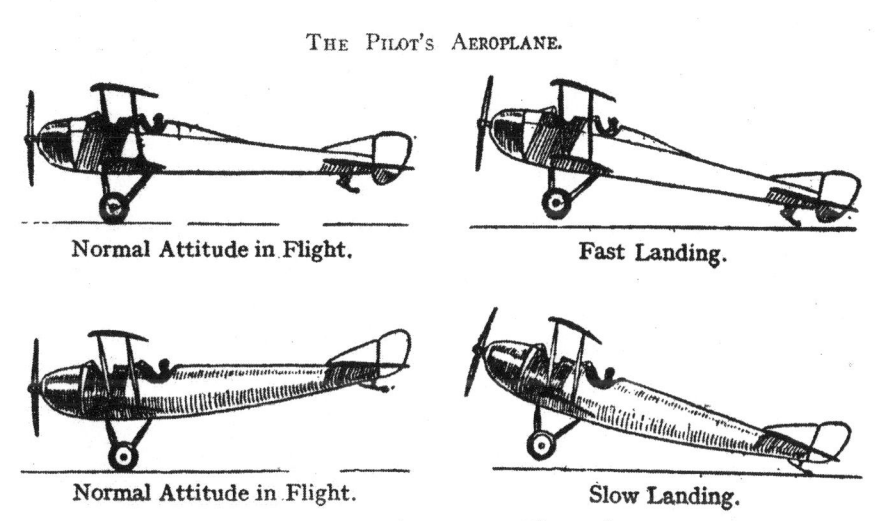

THE PILOT'S AEROPLANE.

Normal Attitude in Flight. **Fast Landing.**

Normal Attitude in Flight. **Slow Landing.**

THE CHANGE OF DESIGN HE WOULD LIKE.

"That's **very** pretty," said the Observer, "but how about Mechanical Difficulties, and Efficiency in respect of Flight? And, anyway, why hasn't such an obvious thing been done already?"

"As regards the first part of your question I assure you that there's nothing in it, and I'll prove it to you as follows——"

"Oh! That's all right, old chap. I'll take your word for it," hurriedly replies the Observer, whose soul isn't tuned to a technical key.

"As regards the latter part of your inquiry," went on the Pilot, a little nettled at having such a poor listener, "it's very simple. Aeroplanes have 'just growed' like Topsy, and they consequently contain this and many another relic of early day design when Aeroplanes were more or less thrown together and anything was good enough that could get off the ground."

"By Jove," interrupts the Observer, "I do believe the fog is lifting. Hadn't we better get the engine and body covers off, just in case it's really so?"

"I believe you're right. I am sure those hills over there could not be seen a few minutes ago, and look—there's sunshine over there. We'd better hurry up."

Ten minutes' hard work and the covers are off, neatly folded and stowed aboard; the picketing wires are cast adrift, and the Pilot is once more in his seat. The Aeroplane has been turned to face the other end of the field, and, the Observer swinging round the propeller, the engine is awake again and slowly ticking over. Quickly the Observer climbs into his seat in front of the Pilot, and, the latter slightly opening the throttle, the Aeroplane leisurely rolls over the ground towards the other end of the field, from which the ascent will be made.

Arriving there the Pilot turns the Aeroplane in order to face the wind and thus secure a quick "get-off." Then he opens the throttle fully and the mighty voice of the Engine roars out "Now see me clear that hedge!" and the Aeroplane races forward at its minimum angle of incidence. Tail up, and with ever-increasing speed, it rushes towards the hedge under the lee of which it has lately been at rest; and then, just as the Observer involuntarily pulls back an imaginary "joy-stick," the Pilot moves the real one and places the machine at its best climbing angle. Like a living thing it responds, and instantly leaves the ground, clearing the hedge like a—well, like an Aeroplane with an excellent margin of lift. Upwards it climbs with even and powerful lift, and the familiar scenes below again gladden the eyes of the Pilot. Smaller and more and more squat grow the houses and hills; more and more doll-like appear the fields which are clearly outlined by the hedges; and soon the country below is easily identified with the map. Now they can see the river before them and a bay of the sea which must be crossed or skirted. The fog still lingers along the course of the river and between the hills, but is fast rolling away in grey, ghost-like masses.

'CROSS COUNTRY

Out to sea it obscures the horizon, making it difficult to be sure where water ends and fog begins, and creating a strange, rather weird effect by which ships at a certain distance appear to be floating in space.

Now the Aeroplane is almost over the river, and the next instant it suddenly drops into a "hole in the air." With great suddenness it happens, and for some two hundred feet it drops nose-down and tilted over sideways; but the Pilot is prepared and has put his craft on an even keel in less time than it takes to tell you about it; for well he knows that he must expect such conditions when passing over a shore or, indeed, any well-defined change in the composition of the earth's surface. Especially is this so on a hot and sunny day, for then the warm surface of the earth creates columns of ascending air, the speed of the ascent depending upon the composition of the surface. Sandy soil, for instance, such as borders this river produces a quickly ascending column of air, whereas water and forests have not such a marked effect. Thus, when our Aeroplane passed over the shore of the river, it suddenly lost the lift due to the ascending air produced by the warm sandy soil, and it consequently dropped just as if it had fallen into a hole.

Now the Aeroplane is over the bay and, the sea being calm, the Pilot looks down, down through the water, and clearly sees the bottom, hundreds of feet below the surface. Down through the reflection of the blue sky and clouds, and one might think that is all, but it isn't. Only those who fly know the beauties of the sea as viewed from above; its dappled pearly tints; its soft dark blue shadows; the beautiful contrasts of unusual shades of colour which are always differing and shifting with the changing sunshine and the ever moving position of the aerial observer. Ah! for some better pen than mine to describe these things! One with glowing words and a magic rhythm to express the wonders of the air and the beauty of the garden beneath—the immensity of the sea—the sense of space and of one's littleness there—the realization of the Power moving the multitudes below—the exaltation of spirit altitude produces—the joy of speed. A new world of sensation!

THE AEROPLANE SPEAKS

Now the bay is almost crossed and the Aerodrome at B can be distinguished.

* * * * * *

On the Aerodrome is a little crowd waiting and watching for the arrival of the Aeroplane, for it is of a new and improved type and its first 'cross-country performance is of keen interest to these men; men who really know something about flight.

There is the Squadron Commander who has done some real flying in his time; several well-seasoned Flight-Commanders; a dozen or more Flight-Lieutenants; a knowledgeable Flight-Sergeant; a number of Air Mechanics, and, a little on one side and almost unnoticed, the Designer.

"I hope they are all right," said someone, "and that they haven't had difficulties with the fog. It rolled up very quickly, you know."

"Never fear," remarked a Flight-Commander. "I know the Pilot well and he's a good 'un; far too good to carry on into a fog."

"They say the machine is really something out of the ordinary," said another, "and that, for once, the Designer has been allowed full play; that he hasn't been forced to unduly standardize ribs, spars, struts, etc., and has more or less had his own way. I wonder who he is. It seems strange we hear so little of him."

"Ah! my boy. You do a bit more flying and you'll discover that things are not always as they appear from a distance!"

"There she is, sir!" cries the Flight-Sergeant. "Just a speck over the silvery corner of that cloud."

A tiny speck it looks, some six miles distant and three thousand feet high; but, racing along, it rapidly appears larger and soon its outlines can be traced and the sunlight be seen playing upon the whirling propeller.

Now the distant drone of the engine can be heard, but not for long, for suddenly it ceases and, the nose of the Aeroplane sinking, the craft commences gliding downwards.

"Surely too far away," says a subaltern. It will be

'CROSS COUNTRY

a wonderful machine if, from that distance and height, it can glide into the Aerodrome." And more than one express the opinion that it cannot be done; but the Designer smiles to himself, yet with a little anxiety, for his reputation is at stake, and Efficiency, the main reward he desires, is perhaps, or perhaps not, at last within his grasp!

Swiftly the machine glides downwards towards them, and it can now be seen how surprisingly little it is affected by the rough weather and gusts; so much so that a little chorus of approval is heard.

"Jolly good gliding angle," says someone; and another, "Beautifully quick controls, what?" and from yet another, "By Jove! The Pilot must be sure of the machine. Look, he's stopped the engine entirely."

Then the Aeroplane with noiseless engine glides over the boundary of the Aerodrome, and, with just a soft soughing sound from the air it cleaves, lands gently not fifty yards from the onlookers.

"Glad to see you," says the Squadron Commander to the Pilot. "How do you like the machine?" And the Pilot replies:

"I never want a better one, sir. It almost flies itself!"

And the Designer turns his face homewards and towards his beloved drawing-office; well satisfied, but still dreaming dreams of the future and . . . looking far ahead whom should he see but Efficiency at last coming towards him! And to him she is all things. In her hair is the morning sunshine; her eyes hold the blue of the sky, and on her cheeks is the pearly tint of the clouds as seen from above. The passion of speed, the lure of space, the sense of power, and the wonder of the future . . . all these things she holds for him.

"Ah!" he cries. "You'll never leave me now, when at last there is no one between us?"

And Efficiency, smiling and blushing, but practical as ever, says:

"And you will never throw those Compromises in my face?"

"My dear, I love you for them! Haven't they been my life ever since I began striving for you ten long years ago?"

THE AEROPLANE SPEAKS

And so they walked off very happily, arm-in-arm together; and if this hasn't bored you and you'd like some more of the same sort of thing, I'd just love to tell you some day of the wonderful things they accomplish together, and of what they dream the future holds in store.

And that's the end of the Prologue.

CHAPTER I

FLIGHT

AIR has weight (about 13 cubic feet = 1 lb.), inertia, and momentum. It therefore obeys Newton's laws* and resists movement. It is that resistance or reaction which makes flight possible.

Flight is secured by driving through the air a surface† inclined upwards and towards the direction of motion.

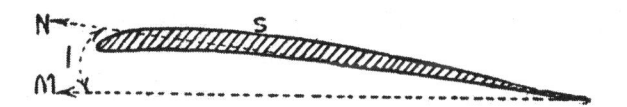

S = Side view of surface.
M = Direction of motion.

CHORD.—The Chord is, for practical purposes, taken to be a straight line from the leading edge of the surface to its trailing edge.

N = A line through the surface starting from its trailing edge. The position of this line, which I call the *Neutral Lift Line*, is found by means of wind-tunnel research, and it varies with differences in the camber (curvature) of surfaces. In order to secure flight, the inclination of the surface must be such that the neutral lift line makes an angle with and *above* the line of motion. If it is coincident with M, there is no lift. If it makes an angle with M and *below* it, then there is a pressure tending to force the surface down.

I = Angle of Incidence. This angle is generally defined as the angle the chord makes with the direction of motion, but that is a bad definition, as it leads to misconception. The angle of incidence is best described as the angle the

---
* See Newton's laws in the Glossary at the end of the book.
† See "Aerofoil" in the Glossary.

neutral lift line makes with the direction of motion relative to the air. You will, however, find that in nearly all rigging specifications the angle of incidence is taken to mean the angle the chord makes with a line parallel to the propeller thrust. This is necessary from the point of view of the practical mechanic who has to rig the aeroplane, for he could not find the neutral lift line, whereas he can easily find the chord. Again, he would certainly be in doubt as to "the direction of motion relative to the air," whereas he can easily find a line parallel to the propeller thrust. It is a pity, however, that these practical considerations have resulted in a bad definition of the angle of incidence becoming prevalent, a consequence of which has been the widespread fallacy that flight may be secured with a negative inclination of the surface. Flight may conceivably be secured with a negative angle of chord, but never with a negative inclination of the surface. All this is only applicable to cambered surfaces. In the case of flat surfaces the neutral lift line coincides with the chord and the definition I have criticised adversely is then applicable. Flat lifting surfaces are, however, never used.

The surface acts upon the air in the following manner:

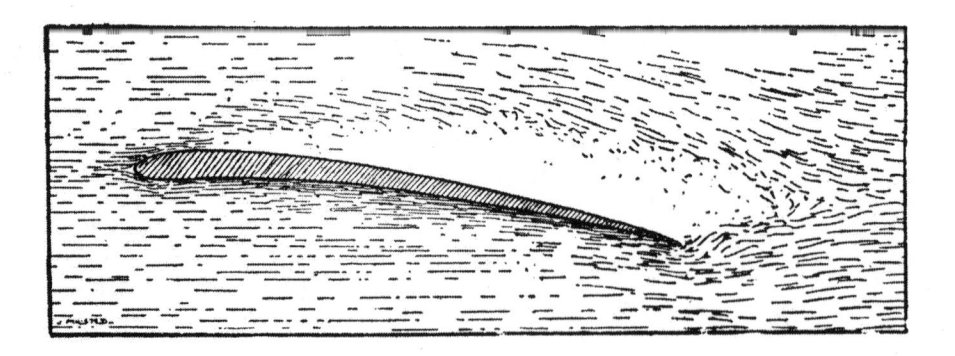

As the bottom of the surface meets the air, it compresses it and accelerates it *downwards*. As a result of this definite action there is, of course, an equal and opposite reaction *upwards*.

The top surface, in moving forward, tends to leave the air behind it, thus creating a semi-vacuum or rarefied area over the top of the surface. Consequently the pressure of air on the top of the surface is decreased, thus assisting the reaction below to lift the surface *upwards*.

The reaction increases approximately as the square of the velocity. It is the result of (1) the mass of air engaged, and (2) the velocity and consequent force with which the surface engages the air. If the reaction was produced by only one of those factors it would increase in direct proportion to the velocity, but, since it is the product of both factors, it increases as V^2.

Approximately three-fifths of the reaction is due to the decrease of density (and consequent decrease of downward pressure) on the top of the surface; and only some two-fifths is due to the upward reaction secured by the action of the bottom surface upon the air. A practical point in respect of this is that, in the event of the fabric covering the surface getting into bad condition, it is more likely to strip off the top than off the bottom.

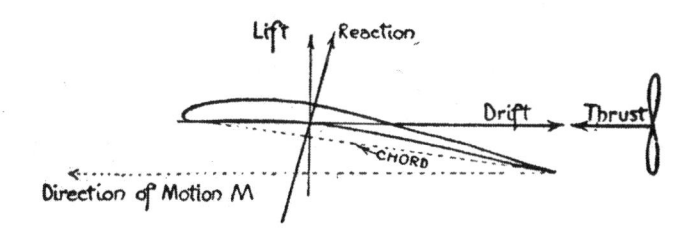

The direction of the reaction is approximately at right-angles to the chord of the surface, as illustrated above; and it is, in considering flight, convenient to divide it into two component parts or values, thus:

1. The vertical component of the reaction, *i.e.*, Lift, which is opposed to Gravity, *i.e.*, the weight of the aeroplane.
2. The horizontal component, *i.e.*, Drift (sometimes called Resistance), to which is opposed the thrust of the propeller.

The direction of the reaction is, of course, the resultant of the forces Lift and Drift.

The Lift is the useful part of the reaction, for it lifts the weight of the aeroplane.

The Drift is the villain of the piece, and must be overcome

by the Thrust in order to secure the necessary velocity to produce the requisite Lift for flight.

DRIFT.—The drift of the whole aeroplane (we have considered only the lifting surface heretofore) may be conveniently divided into three parts, as follows:

Active Drift, which is the drift produced by the lifting surfaces.

Passive Drift, which is the drift produced by all the rest of the aeroplane—the struts, wires, fuselage, under-carriage, etc., all of which is known as "detrimental surface."

Skin Friction, which is the drift produced by the friction of the air with roughnesses of surface. The latter is practically negligible having regard to the smooth surface of the modern aeroplane, and its comparatively slow velocity compared with, for instance, the velocity of a propeller blade.

LIFT-DRIFT RATIO.—The proportion of lift to drift is known as the lift-drift ratio, and is of paramount importance, for it expresses *the efficiency of the aeroplane* (as distinct from engine and propeller). A knowledge of the factors governing the lift-drift ratio is, as will be seen later, *an absolute necessity* to anyone responsible for the rigging of an aeroplane, and the maintenance of it in an efficient and safe condition.

Those factors are as follows:

1. *Velocity.*—The greater the velocity the greater the proportion of drift to lift, and consequently the less the efficiency. Considering the lifting surfaces alone, both the lift and the (active) drift, being component parts of the reaction, increase as the square of the velocity, and the efficiency remains the same at all speeds. But, considering the whole aeroplane, we must remember the passive drift. It also increases as the square of the velocity (with no attendant lift), and, adding itself to the active drift, results in increasing the proportion of total drift (active + passive) to lift.

   But for the increase in passive drift the efficiency

of the aeroplane would not fall with increasing velocity, and it would be possible, by doubling the thrust, to approximately double the speed or lift—a happy state of affairs which can never be, but which we may, in a measure, approach by doing everything possible to diminish the passive drift.

Every effort is then made to decrease it by "stream-lining," *i.e.*, by giving all "detrimental" parts of the aeroplane a form by which they will pass through the air with the least possible drift. Even the wires bracing the aeroplane together are, in many cases, stream-lined, and with a markedly good effect upon the lift-drift ratio. In the case of a certain well-known type of aeroplane the replacing of the ordinary wires by stream-lined wires added over five miles an hour to the flight speed.

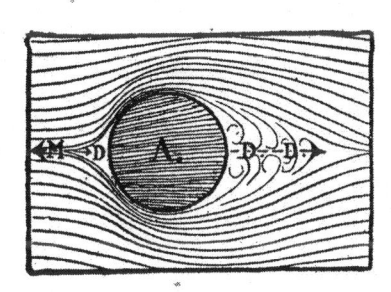

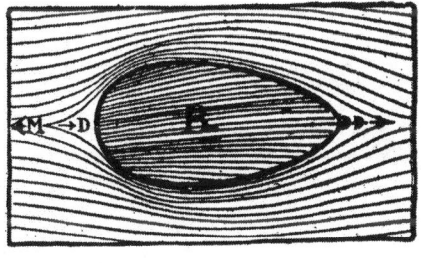

Head-resistance is a term often applied to passive drift, but it is apt to convey a wrong impression, as the drift is not nearly so much the result of the head or forward part of struts, wires, etc., as it is of the rarefied area behind.

Above is illustrated the flow of air round two objects moving in the direction of the arrow M.

In the case of A, you will note that the rarefied area DD is of very considerable extent; whereas in the case of B, the air flows round it in such a way as to meet very closely to the rear of the object, thus *decreasing* DD.

The greater the rarefied area DD, then, the less

the density, and, consequently, the less the pressure of air upon the rear of the object. The less such pressure, then, the better is head-resistance D able to get its work in, and the more thrust will be required to overcome it.

The "fineness" of the stream-line shape, *i.e.*, the proportion of length to width, is determined by the velocity—the greater the velocity, the greater the fineness. The best degree of fineness for any given velocity is found by means of wind-tunnel research.

The practical application of all this is, from a rigging point of view, the importance of adjusting all stream-line parts to be dead-on in the line of flight, but more of that later on.

2. *Angle of Incidence.*—The most efficient angle of incidence varies with the thrust at the disposal of the designer, the weight to be carried, and the climb-velocity ratio desired.

The best angles of incidence for these varying factors are found by means of wind-tunnel research and practical trial and error. Generally speaking, the greater the velocity the smaller should be the angle of incidence, in order to preserve a clean, stream-line shape of rarefied area and freedom from eddies. Should the angle be too great for the velocity, then the rarefied area becomes of irregular shape with attendant turbulent eddies. Such eddies possess no lift value, and since it has taken power to produce them, they represent drift and adversely affect the lift-drift ratio.

From a rigging point of view, one must presume that every standard aeroplane has its lifting surface set at the most efficient angle, and the practical application of all this is in taking the greatest possible care to rig the surface at the correct angle and to maintain it at such angle. Any deviation will adversely affect the lift-drift ratio, *i. e.*, the efficiency.

3. *Camber.*—(Refer to the second illustration in this chapter.) The lifting surfaces are cambered, *i.e.*, curved, in order to decrease the horizontal component of the reaction, *i.e.*, the drift.

The bottom camber: If the bottom of the surface was flat, every particle of air meeting it would do so with a shock, and such shock would produce a very considerable horizontal reaction or drift. By curving it such shock is diminished, and the curve should be such as to produce a uniform (not necessarily constant) acceleration and compression of the air from the leading edge to the trailing edge. Any unevenness in the acceleration and compression of the air produces drift.

The top camber: If this was flat it would produce a rarefied area of irregular shape. I have already explained the bad effect this has upon the lift-drift ratio. The top surface is then curved to produce a rarefied area the shape of which shall be as stream-line and free from attendant eddies as possible.

The camber varies with the angle of incidence, the velocity, and the thickness of the surface. Generally speaking, the greater the velocity, the less the camber and angle of incidence. With infinite velocity the surface would be set at no angle of incidence (the neutral lift line coincident with the direction of motion relative to the air), and would be, top and bottom, of pure stream-line form—*i.e.*, of infinite fineness. This is, of course, carrying theory to absurdity as the surface would then cease to exist.

The best cambers for varying velocities, angles of incidence, and thicknesses of surface, are found by means of wind-tunnel research. The practical application of all this is in taking the greatest care to prevent the surface from becoming distorted and thus spoiling the camber and consequently the lift-drift ratio.

4. *Aspect Ratio.*—This is the proportion of span to

chord. Thus, if the span is, for instance, 50 feet and the chord 5 feet, the surface would be said to have an aspect ratio of 10 to 1.

For *a given velocity* and *a given area* of surface, the greater the aspect ratio, the greater the reaction. It is obvious, I think, that the greater the span, the greater the mass of air engaged, and, as already explained, the reaction is partly the result of the mass of air engaged.

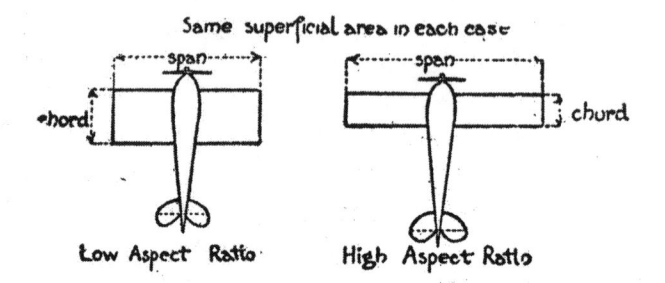

Same superficial area in each case

Low Aspect Ratio   High Aspect Ratio

Not only that, but, *provided* the chord is not decreased to an extent making it impossible to secure the best camber owing to the thickness of the surface, the greater the aspect ratio, the better the lift-drift ratio. The reason of this is rather obscure. It is sometimes advanced that it is owing to the "spill" of air from under the wing-tips. With a high aspect ratio the chord is less than would otherwise be the case. Less chord results in smaller wing-tips and consequently less "spill." This, however, appears to be a rather inadequate reason for the high aspect ratio producing the high lift-drift ratio. Other reasons are also advanced, but they are of such a contentious nature I do not think it well to go into them here. They are of interest to designers, but this is written for the practical pilot and rigger.

5. *Stagger.*—This is the advancement of the top surface relative to the bottom surface, and is not, of course, applicable to a single surface, *i.e.*, a monoplane. In the case of a biplane having no stagger, there will be "interference" and consequent loss of

Efficiency unless the gap between the top and bottom surfaces is equal to not less than 1½ times the chord. If less than that, the air engaged by the bottom of the top surface will have a tendency to be drawn into the rarefied area over the top of the bottom surface, with the result that the surfaces will not secure as good a reaction as would otherwise be the case.

It is not practicable to have a gap of much more than a distance equal to the chord, owing to the drift produced by the great length of struts and wires such a large gap would necessitate. By staggering the top surface forward, however,

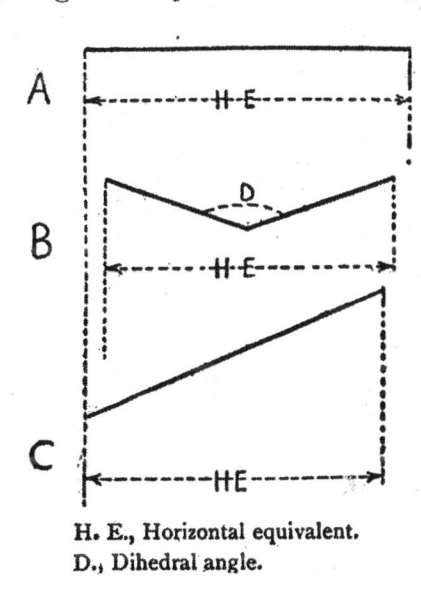

H. E., Horizontal equivalent.
D., Dihedral angle.

it is removed from the action of the lower surface and engages undisturbed air, with the result that the efficiency can in this way be increased by about 5 per cent. Theoretically the top plane should be staggered forward for a distance equal to about 30 per cent. of the chord, the exact distance depending upon the velocity and angle of incidence; but this is not always possible to arrange in designing an aeroplane, owing to difficulties of balance, desired position, and view of pilot, observer, etc.

THE AEROPLANE SPEAKS

6. *Horizontal Equivalent.*—The vertical component of the reaction, *i.e.*, lift, varies as the horizontal equivalent (H.E.) of the surface, but the drift remains the same. Then it follows that if H.E. grows less, the ratio of lift to drift must do the same.

A, B, and C are front views of three surfaces.

A has its full H.E., and therefore, from the point of view from which we are at the moment considering efficiency, it has its best lift-drift ratio.

B and C both possess the same surface as A, but one is inclined upwards from its centre and the other is straight but tilted. For these reasons their H.E.'s are, as illustrated, less than in the case of A. That means less vertical lift, and, the drift remaining the same (for there is the same amount of surface as in A to produce it), the lift-drift ratio falls.

THE MARGIN OF POWER is the power available above that necessary to maintain horizontal flight.

THE MARGIN OF LIFT is the height an aeroplane can gain in a given time and starting from a given altitude. As an example, thus: 1,000 feet the first minute, and starting from an altitude of 500 feet above sea-level.

The margin of lift decreases with altitude, owing to the decrease in the density of the air, which adversely affects the engine. Provided the engine maintained its impulse with altitude, then, if we ignore the problem of the propeller, which I will go into later on, the margin of lift would not disappear. Moreover, greater velocity for a given power would be secured at a greater altitude, owing to the decreased density of air to be overcome. After reading that, you may like to light your pipe and indulge in dreams of the wonderful possibilities which may become realities if some brilliant genius shows us some day how to secure a constant power with increasing altitude. I am afraid, however, that will always remain impossible; but it is probable

FLIGHT

that some very interesting steps may be taken in that direction.

THE MINIMUM ANGLE OF INCIDENCE is the smallest angle at which, for a given power, surface (including detrimental surface), and weight, horizontal flight can be maintained.

THE MAXIMUM ANGLE OF INCIDENCE is the greatest angle at which, for a given power, surface (including detrimental surface), and weight, horizontal flight can be maintained.

THE OPTIMUM ANGLE OF INCIDENCE is the angle at which the lift-drift ratio is highest. In modern aeroplanes it is that angle of incidence possessed by the surface when the axis of the propeller is horizontal.

THE BEST CLIMBING ANGLE is approximately half-way between the maximum and the optimum angles.

All present-day aeroplanes are a compromise between Climb and horizontal Velocity. We will compare the essentials for two aeroplanes, one designed for maximum climb, and the other for maximum velocity.

ESSENTIALS FOR MAXIMUM CLIMB:

1. *Low velocity,* in order to secure the best lift-drift ratio.

2. Having a low velocity, *a large surface* will be necessary in order to engage the necessary mass of air to secure the requisite lift.

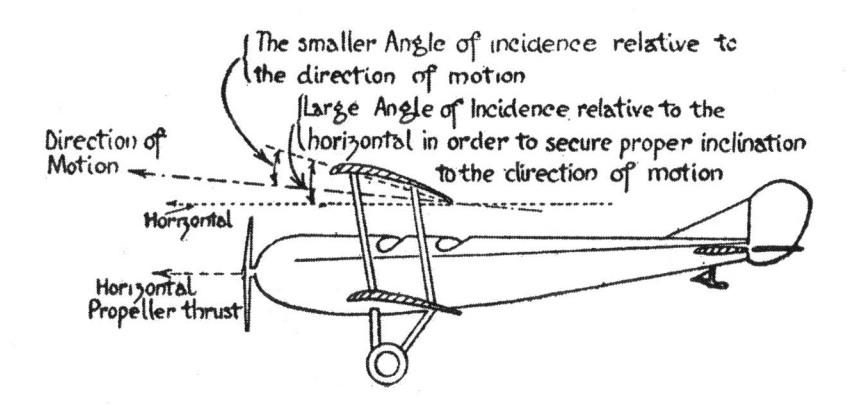

3. Since (1) such a climbing machine will move along an upward sloping path, and (2) will climb with its propeller thrust horizontal, then a *large angle relative to the direction of the thrust* will be necessary in order to secure the requisite angle relative to the direction of motion.

The propeller thrust should be always horizontal, because the most efficient flying-machine (having regard to climb *or* velocity) has, so far, been found to be an arrangement of an inclined surface driven by a *horizontal* thrust—the surface lifting the weight, and the thrust overcoming the drift. This is, in practice, a far more efficient arrangement than the helicopter, *i.e.*, the air-screw revolving about a vertical axis and producing a thrust opposed to gravity. If, when climbing, the propeller thrust is at such an angle as to tend to haul the aeroplane upwards, then it is, in a measure, acting as a helicopter, and that means inefficiency. The reason of a helicopter being inefficient in practice is due to the fact that, owing to mechanical difficulties, it is impossible to construct within a reasonable weight an air-screw of the requisite dimensions. That being so, it would be necessary, in order to absorb the power of the engine, to revolve the comparatively small-surfaced air screw at an immensely greater velocity than that of the aeroplane's surface. As already explained, the lift-drift ratio falls with velocity on account of the increase in passive drift. This applies to a blade of a propeller or air-screw, which is nothing but a revolving surface set at angle of incidence, and which it is impossible to construct without a good deal of detrimental surface near the central boss.

4. The velocity being low, then it follows that for that reason also *the angle of incidence should be comparatively large*.

5. *Camber.*—Since such an aeroplane would be of low velocity, and therefore possess a large angle of incidence, *a large camber* would be necessary.

Let us now consider the essentials for an aeroplane of maximum velocity for its power, and possessing merely enough lift to get off the ground, but no margin of lift.

FLIGHT

1. Comparatively *high velocity*.
2. A comparatively *small surface,* because, being of greater velocity than the maximum climber, a greater mass of air will be engaged for a given surface and time, and therefore a smaller surface will be sufficient to secure the requisit lift.
3. *A small angle relative to the propeller thrust,* since the latter coincides with the direction of motion.
4. A comparatively *small angle of incidence* by reason of the high velocity.
5. A comparatively *small camber* follows as a result of the small angle of incidence.

Summary.

Essentials for Maximum Climb.	*Essentials for Maximum Velocity.*
1. Low velocity.	High velocity.
2. Large surface.	Small surface.
3. Large angle relative to propeller thrust.	Small angle relative to propeller thrust.
4. Large angle relative to direction of motion.	Small angle relative to direction of motion.
5. Large camber.	Small camber.

It is mechanically impossible to construct an aeroplane of reasonable weight of which it would be possible to vary the above opposing essentials. Therefore, all aeroplanes are designed as a compromise between Climb and Velocity.

As a rule aeroplanes are designed to have at low altitude a slight margin of lift when the propeller thrust is horizontal.

ANGLES OF INCIDENCE (INDICATED APPROXIMATELY) OF AN AEROPLANE DESIGNED AS A COMPROMISE BETWEEN VELOCITY AND CLIMB, AND POSSESSING A SLIGHT MARGIN OF LIFT AT A LOW ALTITUDE AND WHEN THE THRUST IS HORIZONTAL.

Minimum Angle.

This gives the greatest velocity during horizontal flight at a low altitude. Greater velocity would be secured if the surface, angle, and camber were smaller and designed to just maintain horizontal flight with a horizontal thrust. Also, in such case, the propeller would not be thrusting downwards, but along a horizontal line which is obviously a more efficient arrangement if we regard the aeroplane merely from one point of view, i.e., either with reference to velocity or climb.

Optimum Angle (Thrust horizontal)

The velocity is less than at the smaller minimum angle, and, as aeroplanes are designed to-day, the area and angle of incidence of the surface is such as to secure a slight ascent at a low altitude. The camber of the surface is designed for this angle of incidence and velocity. The lift-drift ratio is best at this angle.

Best Climbing Angle.

The velocity is now still less by reason of the increased angle producing increase of drift. Less velocity at *a given angle* produces less lift, but the increased angle more or less offsets the loss of lift due to the decreased velocity; and, in addition, the thrust is now hauling the aeroplane upwards.

Maximum Angle.

The greater angle has now produced so much drift as to lessen the velocity to a point where the combined lifts from the surface and from the thrust are only just able to maintain horizontal flight. Any greater angle will result in a still lower lift-drift ratio. The lift will then become less than the weight and the aeroplane will consequently fall. Such a fall is known as "stalling" or "pancaking."

NOTE.—The golden rule for beginners: Never exceed the Best Climbing Angle. Always maintain the flying speed of the aeroplane.

FLIGHT

By this means, when the altitude is reached where the margin of lift disappears (on account of loss of engine power), and which is, consequently, the altitude where it is just possible to maintain horizontal flight, the aeroplane is flying with its thrust horizontal and with maximum efficiency (as distinct from engine and propeller efficiency).

The margin of lift at low altitude, and when the thrust is horizontal, should then be such that the higher altitude at which the margin of lift is lost is that altitude at which most of the aeroplane's horizontal flight work is done. That ensures maximum velocity when most required.

Unfortunately, where aeroplanes designed for fighting are concerned, the altitude where most of the work is done is that at which both maximum velocity and maximum margin of lift for power are required.

Perhaps some day a brilliant inventor will design an aeroplane of reasonable weight and drift of which it will be possible for the pilot to vary at will the above-mentioned opposing essentials. Then we shall get maximum velocity, or maximum margin of lift, for power as required. Until then the design of the aeroplane must remain a compromise between Velocity and Climb.

CHAPTER II

STABILITY AND CONTROL

STABILITY is a condition whereby an object disturbed has a natural tendency to return to its first and normal position. Example: a weight suspended by a cord.

INSTABILITY is a condition whereby an object disturbed has a natural tendency to move as far as possible away from its first position, with no tendency to return. Example: a stick balanced vertically upon your finger.

NEUTRAL INSTABILITY is a condition whereby an object disturbed has no tendency to move farther than displaced by the force of the disturbance, and no tendency to return to its first position.

In order that an aeroplane may be reasonably controllable, it is necessary for it to possess some degree of stability longitudinally, laterally, and directionally.

LONGITUDINAL STABILITY in an aeroplane is its stability about an axis transverse to the direction of normal horizontal flight, and without which it would pitch and toss.

LATERAL STABILITY is its stability about its longitudinal axis, and without which it would roll sideways.

DIRECTIONAL STABILITY is its stability about its vertical axis, and without which it would have no tendency to keep its course.

For such directional stability to exist there must be, in effect,* more "keel-surface" behind the vertical axis than there is in front of it. By keel-surface I mean every-

---

* "In effect" because, although there may be actually the greatest proportion of keel-surface in front of the vertical axis, such surface may be much nearer to the axis than is the keel-surface towards the tail. The latter may then be actually less than the surface in front, but, being farther from the axis, it has a greater leverage, and consequently is greater in effect than the surface in front.

STABILITY AND CONTROL

thing to be seen when looking at an aeroplane from the side of it—the sides of the body, undercarriage, struts, wires, etc. The same thing applies to a weathercock. You know what would happen if there was insufficient keel-surface behind the vertical axis upon which it is pivoted. It would turn off its proper course, which is opposite to the direction of the wind. It is very much the same in the case of an aeroplane.

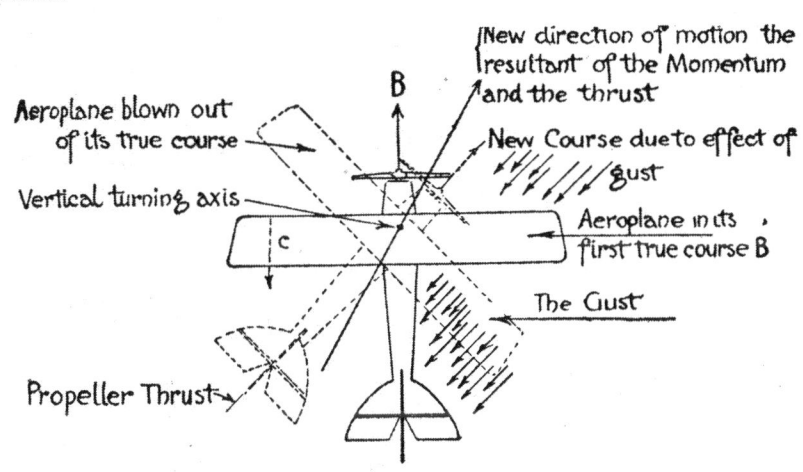

[Diagram: Aeroplane blown out of its true course; Vertical turning axis; Propeller Thrust; B; C; New direction of motion the resultant of the Momentum and the thrust; New Course due to effect of gust; Aeroplane in its first true course B; The Gust]

The above illustration represents an aeroplane (directionally stable) flying along the course B. A gust striking it as indicated acts upon the greater proportion of keel-surface behind the turning axis and throws it into the new course. It does not, however, travel along the new course, owing to its momentum in the direction B. It travels, as long as such momentum lasts, in a direction which is the resultant of the two forces Thrust and Momentum. But the centre line of the aeroplane is pointing in the direction of the new course. Therefore its attitude, relative to the direction of motion, is more or less sideways, and it consequently receives an air prssure in the direction C. Such pressure, acting upon the keel-surface, presses the tail back towards its first position in which the aeroplane is upon its course B.

What I have described is continually going on during flight, but in a well-designed aeroplane such stabilizing movements are, most of the time, so slight as to be imperceptible to the pilot.

If an aeroplane was not stabilized in this way, it would

not only be continually trying to leave its course, but it would also possess a dangerous tendency to "nose away" from the direction of the side gusts. In such case the gust shown in the above illustration would turn the aeroplane round the opposite way a very considerable distance; and the right wing, being on the outside of the turn, would travel with greater velocity than the left wing. Increased velocity means increased lift; and so, the right wing lifting, the aeroplane would turn over sideways very quickly.

LONGITUDINAL STABILITY.—Flat surfaces are longitudinally stable owing to the fact that with decreasing angles of incidence the centre line of pressure (C.P.) moves forward.

The C.P. is a line taken across the surface, transverse to the direction of motion, and about which all the air forces may be said to balance, or through which they may be said to act.

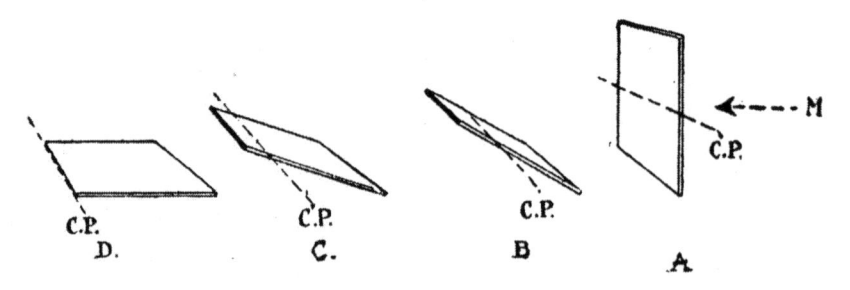

Imagine A to be a flat surface, attitude vertical, travelling through the air in the direction of motion M. Its C.P. is then obviously along the exact centre line of the surface as illustrated.

In B, C, and D the surfaces are shown with angles of incidence decreasing to nothing, and you will note that the C.P. moves forward with the decreasing angle.

Now, should some gust or eddy tend to make the surface decrease the angle, *i.e.*, dive, then the C.P. moves forward and pushes the front of the surface up. Should the surface tend to assume too large an angle, then the reverse happens—the C.P. moves back and pushes the rear of the surface up.

Flat surfaces are, then, theoretically stable longitudinally. They are not, however, used, on account of their poor lift-drift ratio.

STABILITY AND CONTROL

As already explained, cambered surfaces are used, and these are longitudinally unstable at those angles of incidence producing a reasonable lift-drift ratio, *i.e.*, at angles below about 12°.

A is a cambered surface, attitude approximately vertical, moving through the air in the direction M. Obviously the C.P. coincides with the transverse centre line of the surface.

With decreasing angles, down to angles of about 30°, the C.P. moves forward as in the case of flat surfaces (see B), but angles above 30° do not interest us, since they produce a very low ratio of lift to drift.

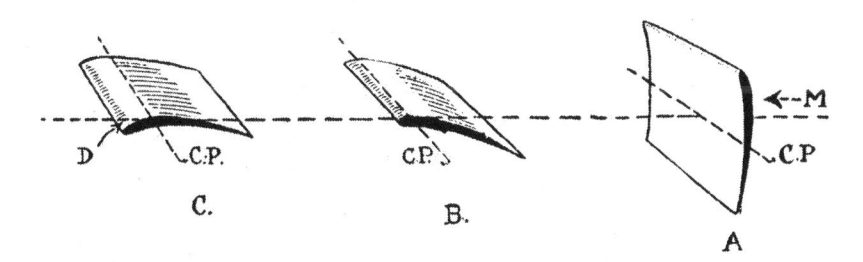

Below angles of about 30° (see C) the dipping front part of the surface assumes a negative angle of incidence resulting in the *downward* air pressure D, and the more the angle of incidence is decreased, the greater such negative angle and its resultant pressure D. Since the C.P. is the resultant of all the air forces, its position is naturally affected by D, which causes it to move backwards. Now, should some gust or eddy tend to make the surface decrease its angle of incidence, *i.e.*, dive, then the C.P. moves backwards, and, pushing up the rear of the surface, causes it to dive the more. Should the surface tend to assume too large an angle, then the reverse happens; the pressure D decreases, with the result that C.P. moves forward and pushes up the front of the surface, thus increasing the angle still further, the final result being a "tail-slide."

It is therefore necessary to find a means of stabilizing the naturally unstable cambered surface. This is usually secured by means of a stabilizing surface fixed some distance in the rear of the main surface, and it is a necessary condition that the neutral lift lines of the two surfaces, when projected

THE AEROPLANE SPEAKS

to meet each other, make a dihedral angle. In other words the rear stabilizing surface must have a lesser angle of incidence than the main surface—certainly not more than one-third of that of the main surface. This is known as the longitudinal dihedral.

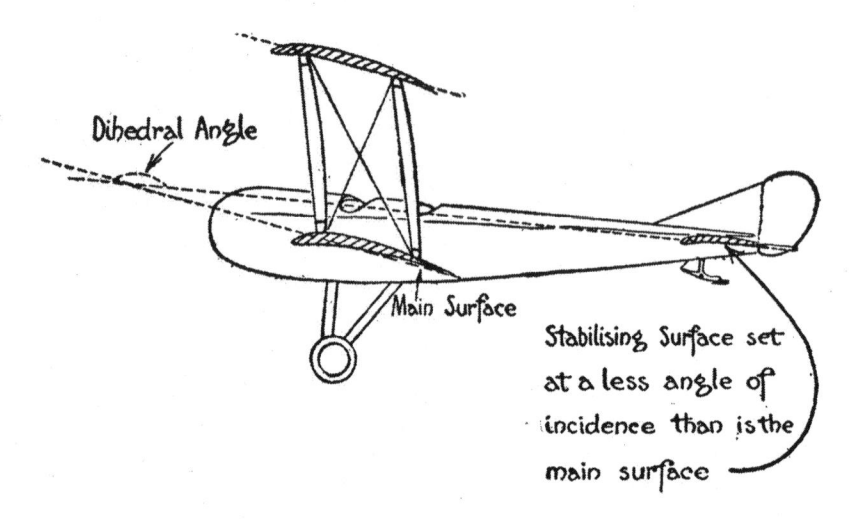

I may add that the tail-plane is sometimes mounted upon the aeroplane at the same angle as the main surface, but, in such cases, it attacks air which has received a downward deflection from the main surface, thus:

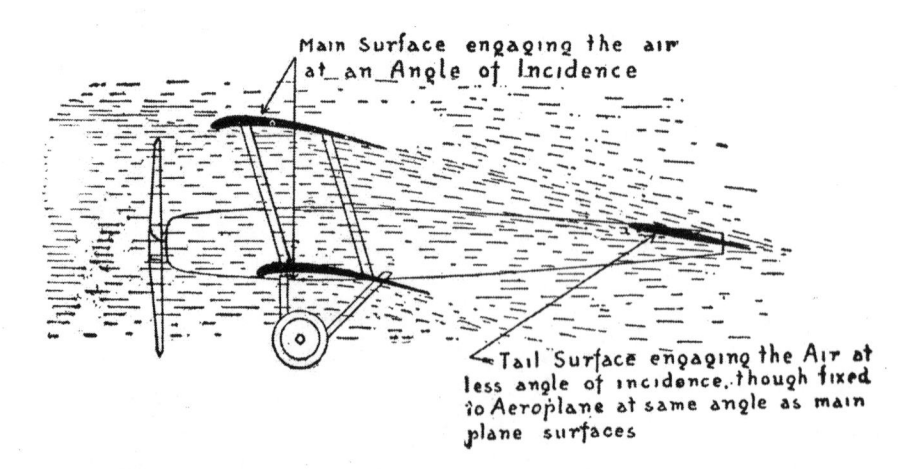

The angle at which the tail surface attacks the air (the angle of incidence) is therefore less than the angle of incidence of the main surface.

STABILITY AND CONTROL

I will now, by means of the following illustration, try to explain how the longitudinal dihedral secures stability:

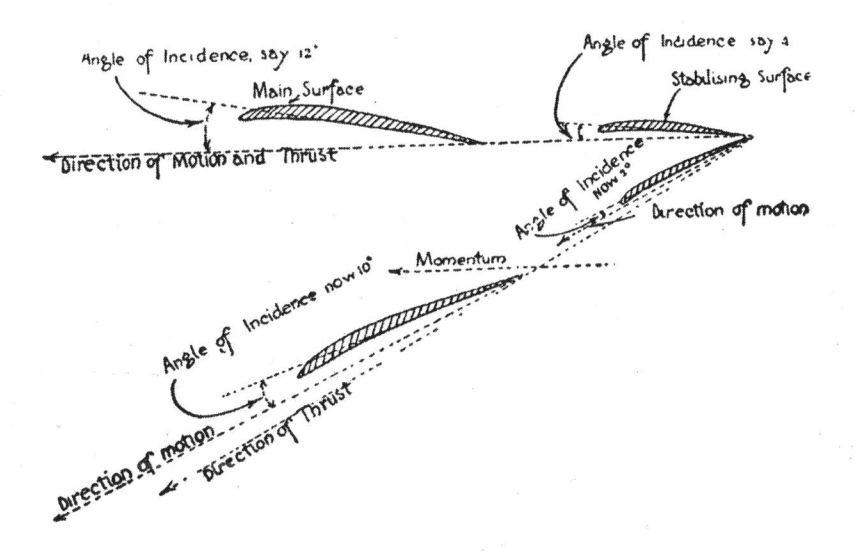

First, imagine the aeroplane travelling in the direction of motion, which coincides with the direction of thrust T. The weight is, of course, balanced about a C.P., the resultant of the C.P. of the main surface and the C.P. of the stabilizing surface. For the sake of illustration, the stabilizing surface has been given an angle of incidence, and therefore has a lift and C.P. In practice the stabilizer is often set at no angle of incidence. In such case the proposition remains the same, but it is, perhaps, a little easier to illustrate it as above.

Now, we will suppose that a gust or eddy throws the machine into the lower position. It no longer travels in the direction of T, since the momentum in the old direction pulls it off that course. M is now the resultant of the Thrust and the Momentum, and you will note that this results in a decrease in the angle our old friend the neutral lift line makes with M, *i.e.*, a decrease in the angle of incidence and therefore a decrease in lift.

We will suppose that this decrease is 2°. Such decrease applies to both main surface and stabilizer, since both are fixed rigidly to the aeroplane.

The main surface, which had 12° angle, has now only 10°, *i.e.*, a loss of *one-sixth*.

THE AEROPLANE SPEAKS

The stabilizer, which had 4° angle, has now only 2°, *i.e.*, a loss of *one-half*.

The latter has therefore lost a greater *proportion* of its angle of incidence, and consequently its lift, than has the main surface. It must then fall relative to the main surface. The tail falling, the aeroplane then assumes its first position, though at a slightly less altitude.

Should a gust throw the nose of the aeroplane up, then the reverse happens. Both main surface and stabilizer increase their angles of incidence in the same amount, but the angle, and therefore the lift, of the stabilizer increases in greater proportion than does the lift of the main surface, with the result that it lifts the tail. The aeroplane then assumes its first position, though at a slightly greater altitude.

Do not fall into the widespread error that the angle of incidence varies as the angle of the aeroplane to the horizontal. It varies with such angle, but not as anything approaching it. Remember that the stabilizing effect of the longitudinal dihedral lasts only as long as there is momentum in the direction of the first course.

These stabilizing movements are taking place all the time, even though imperceptible to the pilot.

Aeroplanes have, in the past, been built with a stabilizing surface in front of the main surface instead of at the rear of it. In such design the main surface (which is then the tail surface as well as the principal lifting surface) must be set at a less angle than the forward stabilizing surface, in order to secure a longitudinal dihedral. The defect of such design lies in the fact that the main surface must have a certain angle to lift the weight—say 5°. Then, in order to secure a sufficiency of longitudinal stability, it is necessary to set the forward stabilizer at about 15°. Such a large angle of incidence results in a very poor lift-drift ratio (and consequently great loss of efficiency), except at very low velocities compared with the speed of modern aeroplanes. At the time such aeroplanes were built velocities were comparatively low, and this defect was, for that reason, not sufficiently appreciated. In the end it killed the "canard" or "tail-first" design.

STABILITY AND CONTROL

Aeroplanes of the Dunne and similar types possess no stabilizing surface distinct from the main surface, but they have a longitudinal dihedral which renders them stable.

The main surface towards the wing-tips is given a decreasing angle of incidence and corresponding camber. The wing-tips then act as longitudinal stabilizers.

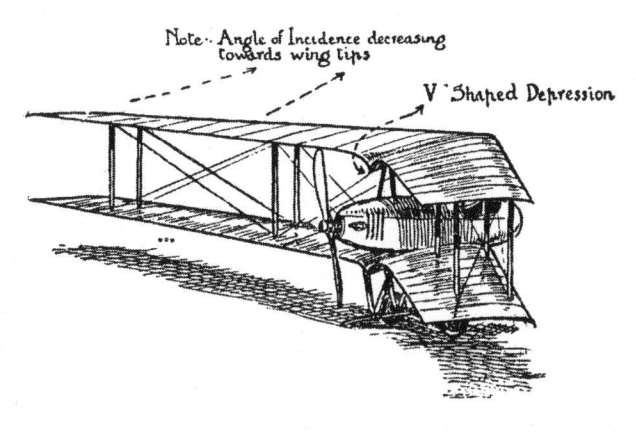

This design of aeroplane, while very interesting, has not proved very practicable, owing to the following disadvantages: (1) The plan design is not, from a mechanical point of view, so sound as that of the ordinary aeroplane surface, which is, in plan, a parallelogram. It is, then, necessary to make the strength of construction greater than would otherwise be the case. That means extra weight. (2) The plan of the surface area is such that the aspect ratio is not so high as if the surface was arranged with its leading edges at right angles to the direction of motion. The lower the aspect ratio, then, the less the lift. This design, then, produces less lift for weight of surface than would the same surface if arranged as a parallelogram. (3) In order to secure the longitudinal dihedral, the angle of incidence has to be very much decreased towards the wing-tips. Then, in order that the lift-drift ratio may be preserved, there must be a corresponding decrease in the camber. That calls for surface ribs of varying cambers, and results in an expensive and lengthy job for the builder. (4) In order to secure directional stability, the surface is, in the centre, arranged to dip down in the form of a V, pointing towards the direction of motion. Should the aeroplane turn off its course, then its momentum in the direction of its first course causes it to move in a direction the resultant of the thrust and the momentum. It then moves in a more or less sideways attitude, which results in an air pressure upon one side of the V, and which tends to turn the aeroplane back to its first course. This arrangement of the surface results in a bad drift. Vertical surfaces at the wing-tips may also be set at an angle producing the same stabilizing effect, but they also increase the drift.

The gyroscopic action of a rotary engine will affect the longitudinal stability when an aeroplane is turned to right or left. In the case of a Gnome engine, such gyroscopic action will tend to depress the nose of the aeroplane when it is turned to the left, and to elevate it when it is turned to the right. In modern aeroplanes this tendency is not sufficiently important to bother about. In the old days of crudely designed and under-powered aeroplanes this gyroscopic action was very marked, and led the majority of pilots to dislike turning an aeroplane to the right, since, in doing so, there was some danger of "stalling."

LATERAL STABILITY is far more difficult for the designer to secure than is longitudinal or directional stability. Some degree of lateral stability may be secured by means of the "lateral dihedral," *i.e.*, the upward inclination of the surface towards its wing-tips thus:

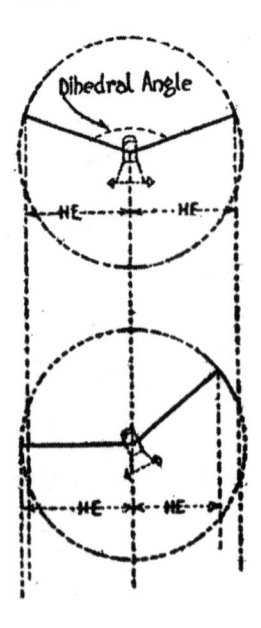

STABILITY AND CONTROL

Imagine the top **V**, illustrated opposite, to be the front view of a surface flying towards you. The horizontal equivalent (H.E.) of the left wing is the same as that of the right wing. Therefore, the lift of one wing is equal to the lift of the other, and the weight, being situated always in the centre, is balanced.

If some movement of the air causes the surface to tilt sideways, as in the lower illustration, then you will note that the H.E. of the left wing increases, and the H.E. of the right wing decreases. The left wing then, having the greatest lift, rises; and the surface assumes its first and normal position.

Unfortunately however, the righting effect is not proportional to the difference between the right and left H.E.'s.

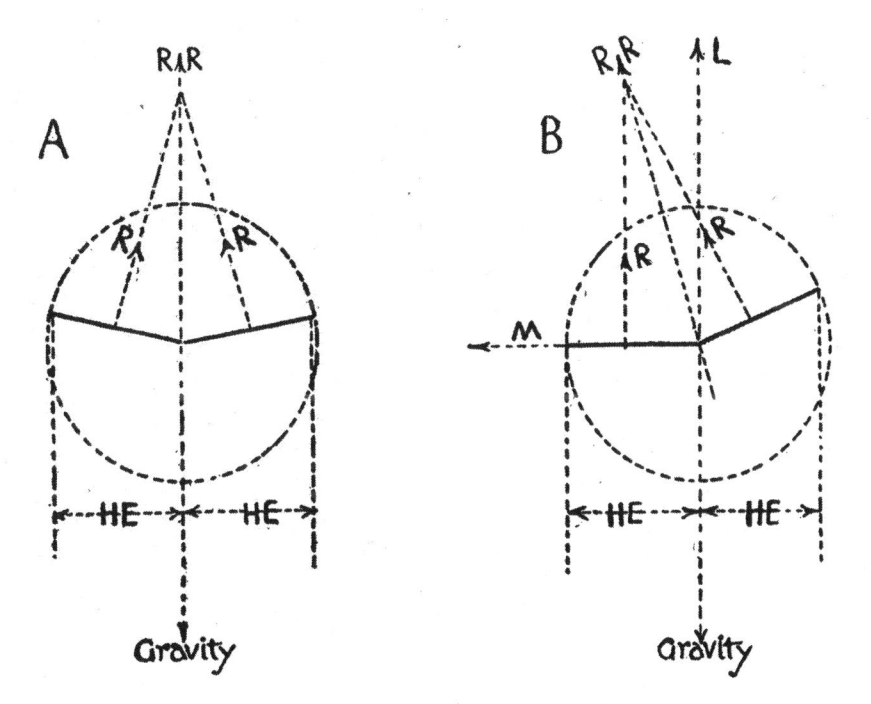

R, Direction of reaction of wing indicated.
R R, Resultant direction of reaction of both wings.
M, Horizontal (sideway) component of reaction.
L, Vertical component of reaction (lift).

In the case of A, the resultant direction of the reaction of both wings is opposed to the direction of gravity or weight. The two forces R R and gravity are then evenly balanced, and the surface is in a state of equilibrium.

THE AEROPLANE SPEAKS

In the case of B, you will note that the R R is not directly opposed to gravity. This results in the appearance of M, and so the resultant direction of motion of the aeroplane is no longer directly forward, but is along a line the resultant of the thrust and M. In other words, it is, while flying forward, at the same time moving sideways in the direction M.

In moving sideways, the keel-surface receives, of course, a pressure from the air equal and opposite to M. Since such surface is greatest in effect towards the tail, then the latter must be pushed sideways. That causes the aeroplane to turn; and, the highest wing being on the outside of the turn, it has a greater velocity than the lower wing. That produces greater lift, and tends to tilt the aeroplane over still more. Such tilting tendency is, however, opposed by the difference in the H.E.'s of the two wings.

It then follows that, for the lateral dihedral angle to be effective, such angle must be large enough to produce, when the aeroplane tilts, a difference in the H.E.'s of the two wings, which difference must be sufficient to not only oppose the tilting tendency due to the aeroplane turning, but sufficient to also force the areoplane back to its original position of equilibrium.

It is now, I hope, clear to the reader that the lateral dihedral is not quite so effective as would appear at first sight. Some designers, indeed, prefer not to use it, since its effect is not very great, and since it must be paid for in loss of H.E. and consequently loss of lift, thus decreasing the lift-drift ratio, *i.e.*, the efficiency. Also, it is sometimes advanced that the lateral dihedral increases the "spill" of air from the wing-tips and that this adversely affects the lift-drift ratio.

The disposition of the keel-surface affects the lateral stability. It should be, in effect, equally divided by the longitudinal turning axis of the aeroplane. If there is an excess of keel-surface above or below such axis, then a side gust striking it will tend to turn the aeroplane over sideways.

The position of the centre of gravity affects lateral stability. If too low, it produces a pendulum effect and causes the aeroplane to roll sideways.

If too high, it acts as a stick balanced vertically would act. If disturbed, it tends to travel to a position as far as

STABILITY AND CONTROL

possible from its original position. It would then tend, when moved, to turn the aeroplane over sideways and into an upside-down position.

From the point of view of lateral stability, the best position for the centre of gravity is one a little below the centre of drift.

Propeller torque affects lateral stability. An aeroplane tends to turn over sideways in the opposite direction to which the propeller revolves.

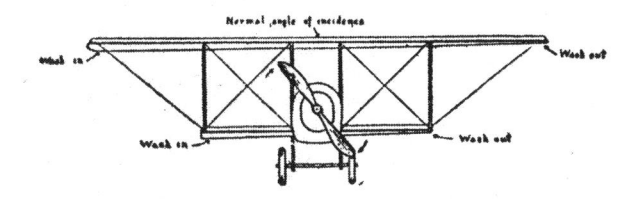

This tendency is offset by increasing the angle of incidence (and consequently the lift) of the side tending to fall; and it is always advisable, if practical considerations allow it, to also decrease the angle upon the other side. In that way it is not necessary to depart so far from the normal angle of incidence at which the lift-drift ratio is highest.

Wash-in is the term applied to the increased angle.

Wash-out is the term applied to the decreased angle.

Both lateral and directional stability may be improved by washing out the angle of incidence on both sides of the surface, thus:

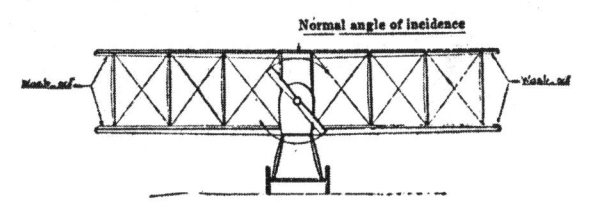

The decreased angle decreases the drift and therefore the effect of gusts upon the wing-tips which is just where they have the most effect upon the aeroplane, owing to the distance from the turning axis.

The wash-out also renders the ailerons (lateral controlling services) more effective, as, in order to operate them, it is

THE AEROPLANE SPEAKS

not then necessary to give them such a large angle of incidence as would otherwise be required.

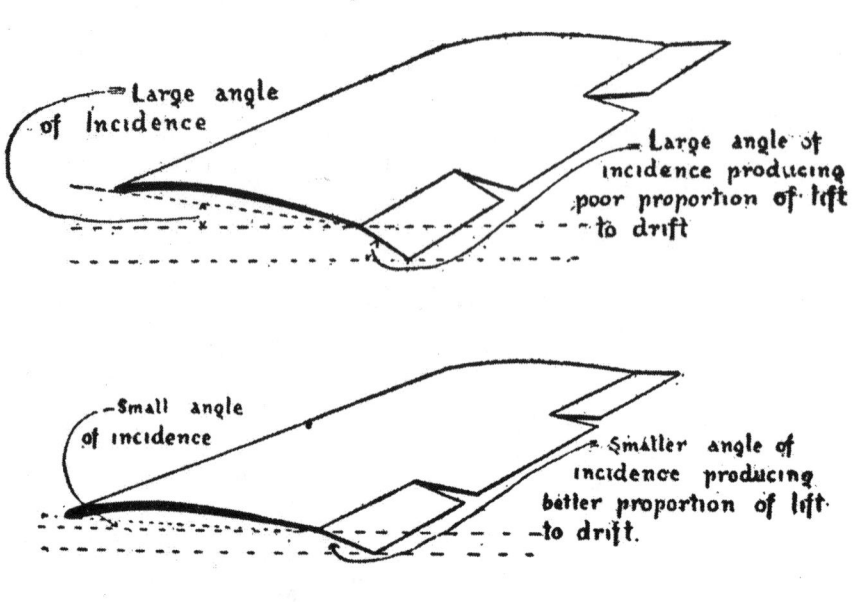

Large angle of incidence

Large angle of incidence producing poor proportion of lift to drift

Small angle of incidence

Smaller angle of incidence producing better proportion of lift to drift.

NOTE Observe that the inclination of the ailerons to the surface is the same in each case

The less the angle of incidence of the ailerons, the better their lift-drift ratio, *i.e.*, their efficiency. You will note that, while the aileron attached to the surface with washed-out angle is operated to the same extent as the aileron illustrated above it, its angle of incidence is considerably less. Its efficiency is therefore greater.

The advantages of the wash-in must, of course, be paid for in some loss of lift, as the lift decreases with the decreased angle.

In order to secure all the above described advantages, a combination is sometimes effected, thus:

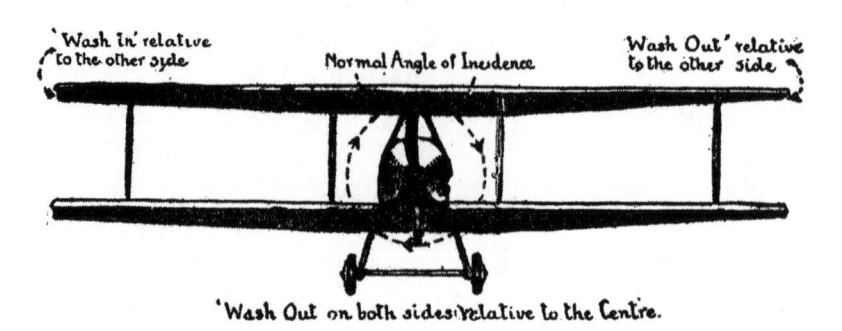

'Wash in' relative to the other side — *Normal Angle of Incidence* — *'Wash Out' relative to the other side*

'Wash Out on both sides' relative to the Centre.

STABILITY AND CONTROL

Banking.—An aeroplane turned off its course to right or left does not at once proceed along its new course. Its momentum in the direction of its first course causes it to travel along a line the resultant of such momentum and the thrust. In other words, it more or less skids sideways and away from the centre of the turn. Its lifting surfaces do not then meet the air in their correct attitude, and the lift may fall to such an extent as to become less than the weight, in which case the aeroplane must fall. This bad effect is minimized by "banking," *i.e.*, tilting the aeroplane sideways. The bottom of the lifting surface is in that way opposed to the air through which it is moving in the direction of the momentum and receives an opposite air pressure. The rarefied area over the top of the surface is rendered still more rare, and this, of course, assists the air pressure in opposing the momentum.

The velocity of the "skid," or sideways movement, is then only such as is necessary to secure an air pressure equal and opposite to the centrifugal force of the turn.

The sharper the turn, the greater the effect of the centrifugal force, and therefore the steeper should be the "bank." *Experentia docet.*

The position of the centre of gravity affects banking. A low C.G. will tend to swing outward from the centre of the turn, and will cause the aeroplane to bank—perhaps too much, in which case the pilot must remedy matters by operating the ailerons.

A high C.G. also tends to swing outward from the centre of the turn. It will tend to make the aeroplane bank the wrong way, and such effect must be remedied by means of the ailerons.

The pleasantest machine from a banking point of view is one in which the C.G. is a little below the centre of drift. It tends to bank the aeroplane the right way for the turn, and the pilot can, if necessary, perfect the bank by means of the ailerons.

The disposition of the keel-surface affects banking. It should be, in effect, evenly divided by the longitudinal axis. An excess of keel-surface above the longitudinal axis will, when banking, receive an air pressure causing the aeroplane to bank, perhaps too much. An excess of keel-surface below the axis has the reverse effect.

SIDE-SLIPPING.—This usually occurs as a result of over-banking. It is always the result of the aeroplane tilting sideways and thus decreasing the horizontal equivalent, and therefore the lift, of the surface. An excessive "bank," or sideways tilt, results in the H.E., and therefore the lift, becoming less than the weight, when, of course, the aeroplane must fall, *i.e.*, side-slip.

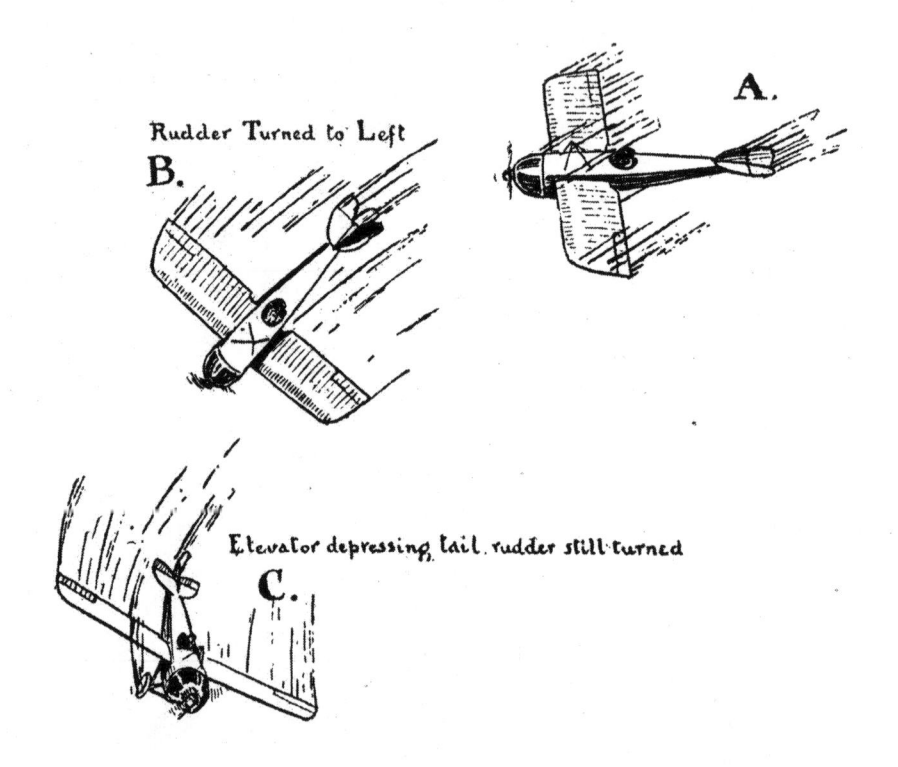

When making a very sharp turn it is necessary to bank very steeply indeed. If, at the same time, the longitudinal axis of the aeroplane remains approximately horizontal, then there must be a fall, and the direction of motion will be the resultant of the thrust and the fall as illustrated above in sketch A. The lifting surfaces and the controlling surfaces are not then meeting the air in the correct attitude, with the result that, in addition to falling, the aeroplane will probably become quite unmanageable.

STABILITY AND CONTROL

The Pilot, however, prevents such a state of affairs from happening by "nosing-down," *i.e.,* by operating the rudder to turn the nose of the aeroplane downward and towards the direction of motion as illustrated in sketch B. This results in the higher wing, which is on the outside of the turn, travelling with greater velocity, and therefore securing a greater reaction than the lower wing, thus tending to tilt the aeroplane over still more. The aeroplane is now almost upside-down, *but* its attitude relative to the direction of motion is correct and the controlling surfaces are all of them working efficiently. The recovery of a normal attitude relative to the Earth is then made as illustrated in sketch C.

The Pilot must then learn to know just the angle of bank at which the margin of lift is lost, and, if a sharp turn necessitates banking beyond that angle, he must "nose-down."

In this matter of banking and nosing-down, and, indeed, regarding stability and control generally, the golden rule for all but very experienced pilots should be: <u>Keep the aeroplane in such an attitude that the air pressure is always directly in the pilot's face.</u> The aeroplane is then always engaging the air as designed to do so, and both lifting and controlling surfaces are acting efficiently. The only exception to this rule is a vertical dive, and I think that is obviously not an attitude for any but very experienced pilots to hanker after.

SPINNING.—This is the worst of all predicaments the pilot can find himself in. Fortunately it rarely happens.

It is due to the combination of (1) a very steep spiral descent of small radius, and (2) insufficiency of keel-surface behind the vertical axis, or the jamming of the rudder end or elevator into a position by which the aeroplane is forced into an increasingly steep and small spiral.

Owing to the small radius of such a spiral, the mass of the aeroplane may gain a rotary momentum greater, in effect, than the air pressure of the keel-surface or controlling surfaces opposed to it; and, when once such a condition occurs, it is difficult to see what can be done by the pilot to remedy it. The sensible pilot will not go beyond reasonable limits of steepness and radius when executing spiral descents.

THE AEROPLANE SPEAKS

GLIDING DESCENT WITHOUT PROPELLER THRUST.—All aeroplanes are, or should be, designed to assume their gliding angle when the power and thrust is cut off. This relieves the pilot of work, worry, and danger should he find himself

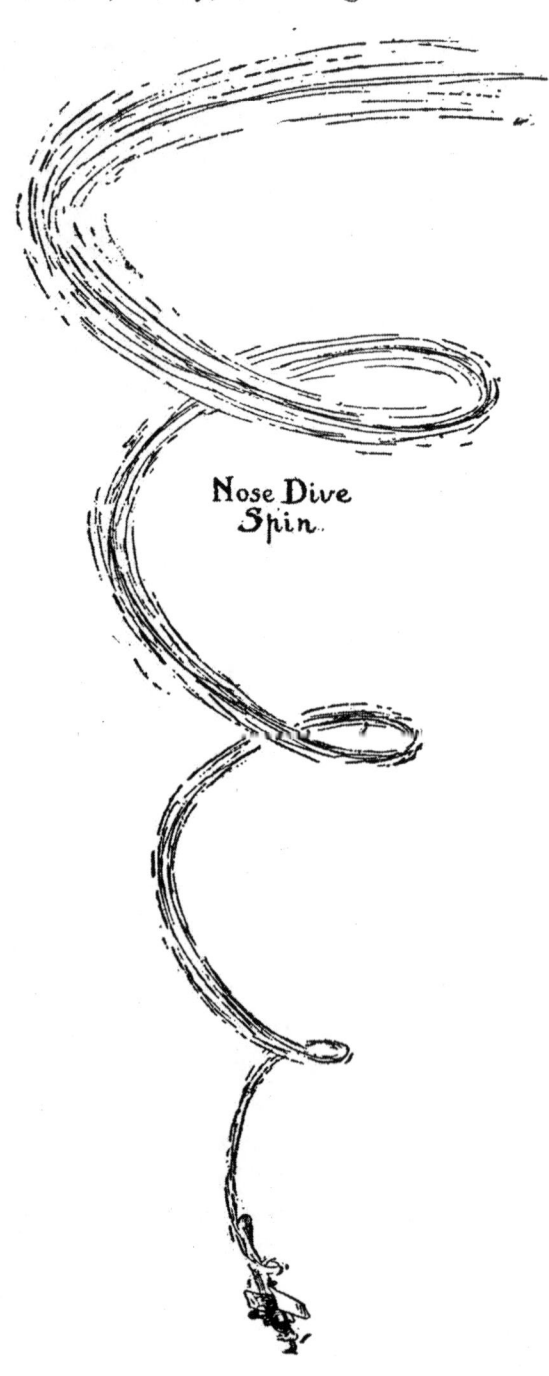

Nose Dive Spin.

STABILITY AND CONTROL

in a fog or cloud. The Pilot, although he may not realize it, maintains the correct attitude of the aeroplane by observing its position relative to the horizon. Flying into a fog or cloud the horizon is lost to view, and he must then rely upon his instruments—(1) the compass for direction; (2) an inclinometer (arched spirit-level) mounted transversely to the longitudinal axis, for lateral stability; and (3) an inclinometer mounted parallel to the longitudinal axis, or the air-speed indicator, which will indicate a nose-down position by increase in air speed, and a tail-down position by decrease in air speed.

The pilot is then under the necessity of watching three instruments and manipulating his three controls to keep the instruments indicating longitudinal, lateral, and directional stability. That is a feat beyond the capacity of the ordinary man. If, however, by the simple movement of throttling down the power and thrust, he can be relieved of looking after the longitudinal stability, he then has only two instruments to watch. That is no small job in itself, but it is, at any rate, fairly practicable.

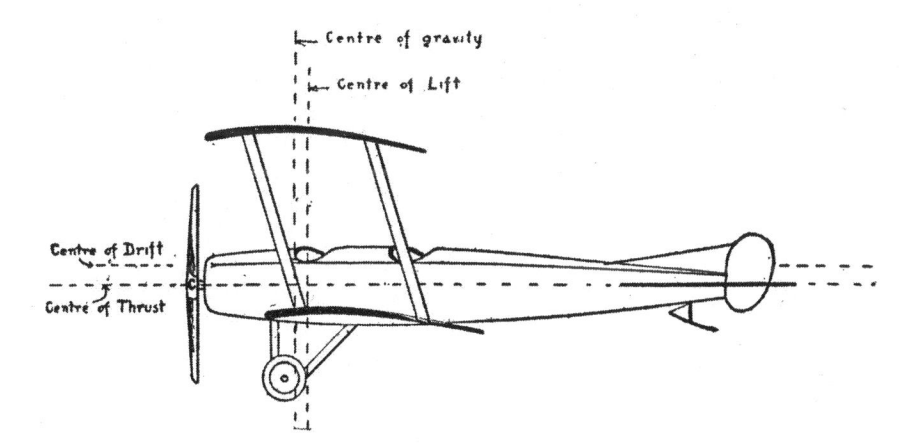

Aeroplanes are, then, designed, or should be, so that the centre of gravity is slightly forward of centre of lift. The aeroplane is then, as a glider, nose-heavy—and the distance the C.G. is placed in advance of the C.L. should be such as to ensure a gliding angle producing a velocity the same as the normal flying speed (for which the strength of construction has been designed).

THE AEROPLANE SPEAKS

In order that this nose-heavy tendency should not exist when the thrust is working and descent not required, the centre of thrust is placed a little below the centre of drift or resistance, and thus tends to pull up the nose of the aeroplane.

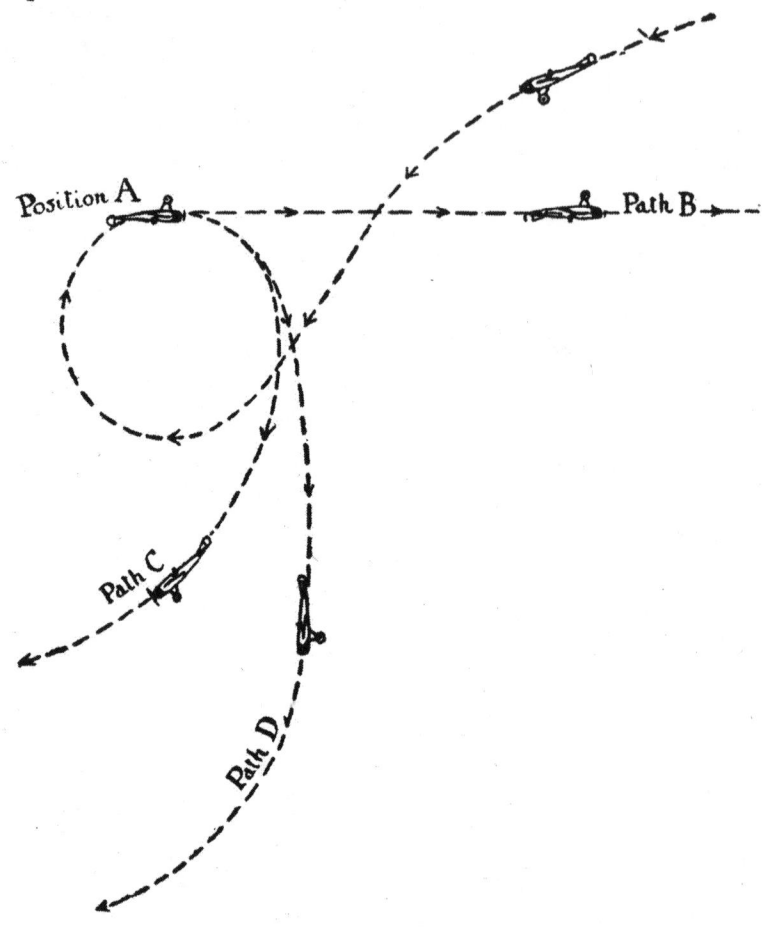

The distance the centre of thrust is placed below the centre of drift should be such as to produce a force equal and opposite to that due to the C.G. being forward of the C.L. (see illustration on p. 87).

LOOPING AND UPSIDE-DOWN FLYING.—If a loop is desired, it is best to throttle the engine down at point A. The C.G. being forward of the C.P., then causes the aeroplane to nose-down, and assists the pilot in making a reasonably small loop along the course C and in securing a quick recovery. If the engine is not throttled down, then the aeroplane may

STABILITY AND CONTROL

be expected to follow the course D, which results in a longer nose dive than in the case of the course C.

A steady, gentle movement of the elevator is necessary. A jerky movement may change the direction of motion so suddenly as to produce dangerous air stresses upon the surfaces, in which case there is a possibility of collapse.

If an upside-down flight is desired, the engine may, or may not, be throttled down at point A. If not throttled down, then the elevator must be operated to secure a course approximately in the direction B. If it is throttled down, then the course must be one of a steeper angle than B, or there will be danger of stalling.

Diagram p. 88.—This is not set at quite the correct angle. Path B should slope slightly downwards from Position A.

CHAPTER III

RIGGING

IN order to rig an aeroplane intelligently, and to maintain it in an efficient and safe condition, it is necessary to possess a knowledge of the stresses it is called upon to endure, and the strains likely to appear.

STRESS is the load or burden a body is called upon to bear. It is usually expressed by the result found by dividing the load by the number of superficial square inches contained in the cross-sectional area of the body.

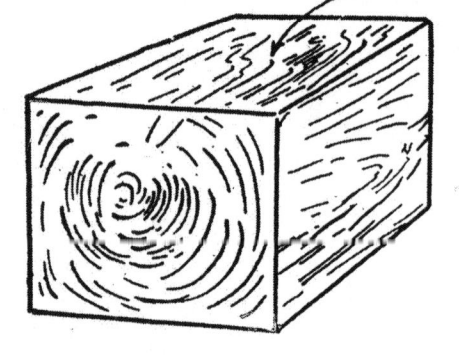

Thus, if, for instance, the object illustrated above contains 4 square inches of cross-sectional area, and the total load it is called upon to endure is 10 tons, the stress would be expressed as 2½ tons.

STRAIN is the deformation produced by stress.

THE FACTOR OF SAFETY is usually expressed by the result found by dividing the stress at which it is known the body will collapse, by the maximum stress it will be called upon to endure. For instance, if a control wire be called upon to endure a maximum stress of 2 cwts., and the known stress

RIGGING

at which it will collapse is 10 cwts., the factor of safety is then 5.

COMPRESSION.—The simple stress of compression tends to produce a crushing strain. Example: the interplane and fuselage struts.

TENSION.—The simple stress of tension tends to produce the strain of elongation. Example: all the wires.

BENDING.—The compound stress of bending is a combination of compression and tension.

The above sketch illustrates a straight piece of wood of which the top, centre, and bottom lines are of equal length. We will now imagine it bent to form a circle, thus:

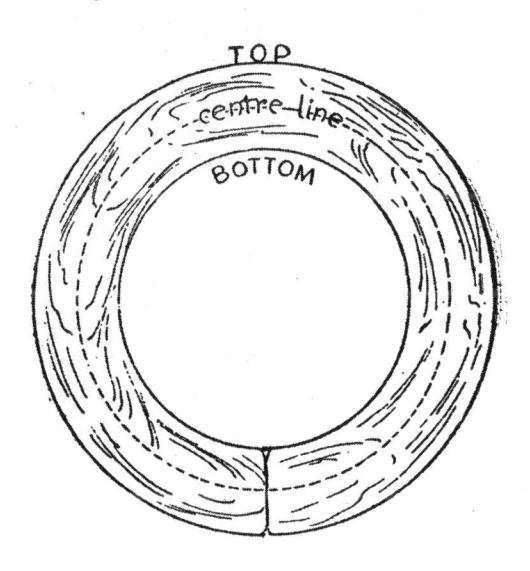

The centre line is still the same length as before being bent; but the top line, being farther from the centre of the circle, is now longer than the centre line. That can be due only to the strain of elongation produced by the stress of tension. The wood between the centre line and the top line is then in tension; and the farther from the centre, the greater the strain, and consequently the greater the tension.

The bottom line, being nearest to the centre of the circle, is now shorter than the centre line. That can be due only to the strain of crushing produced by the stress of compression. The wood between the centre and bottom lines is then in compression; and the nearer the centre of the circle, the greater the strain, and consequently the greater the compression.

It then follows that there is neither tension nor compression, *i.e.*, no stress, at the centre line, and that the wood immediately surrounding it is under considerably less stress than the wood farther away. This being so, the wood in the centre may be hollowed out without unduly weakening struts and spars. In this way 25 to 33 per cent. is saved in the weight of wood in an aeroplane.

The strength of wood is in its fibres, which should, as far as possible, run without break from one end of a strut or spar to the other end. A point to remember is that the outside fibres, being farthest removed from the centre line, are doing by far the greatest work.

SHEAR STRESS is such that, when material collapses under it, one part slides over the other. Example: all the locking pins.

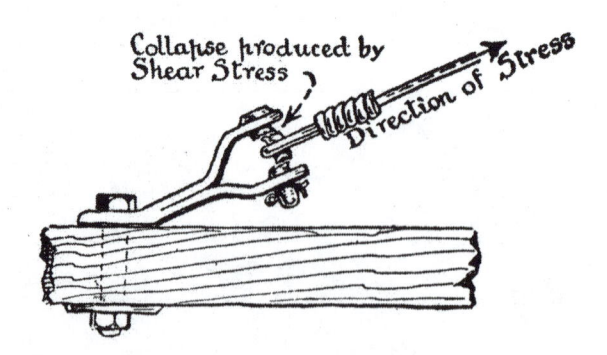

Some of the bolts are also in shear or "sideways" stress, owing to lugs under their heads and from which wires are taken. Such a wire, exerting a sideways pull upon a bolt, tries to break it in such a way as to make one piece of the bolt slide over the other piece.

RIGGING

Torsion.—This is a twisting stress compounded of compression, tension, and shear stresses. Example: the propeller shaft.

Nature of Wood Under Stress.—Wood, for its weight, takes the stress of compression far better than any other stress. For instance: a walking-stick of less than 1 lb. in weight will, if kept perfectly straight, probably stand up to a compression stress of a ton or more before crushing; whereas, if the same stick is put under a bending stress, it will probably collapse to a stress of not more than about 50 lb. That is a very great difference, and, since weight is of the greatest importance, the design of an aeroplane is always such as to, as far as possible, keep the various wooden parts of its construction in direct compression. Weight being of such vital importance, and designers all trying to outdo each other in saving weight, it follows that the factor of safety is rather low in an aeroplane. The parts in direct compression will, however, take the stresses safely provided the following conditions are carefully observed.

Conditions to be Observed:

1, *All the spars and struts must be perfectly straight.*

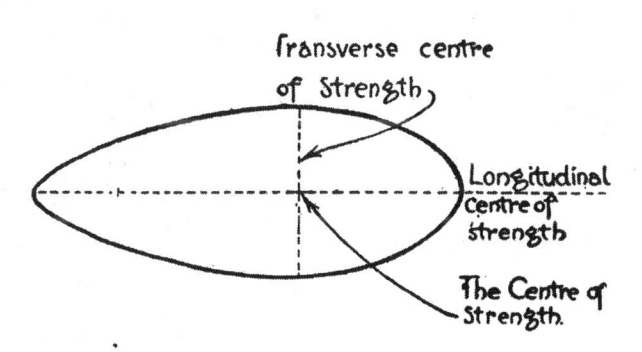

The above sketch illustrates a section through an interplane strut. If the strut is to be kept straight, *i.e.*, prevented from bending, then the stress of compression must be equally disposed about the centre of strength. If it is not straight, then there will be more compression on one side of the centre of strength than on the other side. That is a step towards getting compression on one side

THE AEROPLANE SPEAKS

and tension on the other side, in which case it may be forced to take a bending stress for which it is not designed. Even if it does not collapse it will, in effect, become shorter, and thus throw out of adjustment the gap and all the wires attached to the top and bottom of the strut, with the result that the flight efficiency of the aeroplane will be spoiled.

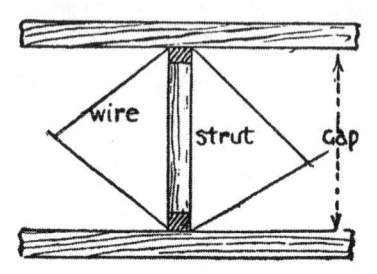

Strut straight. Wires and gap correctly adjusted

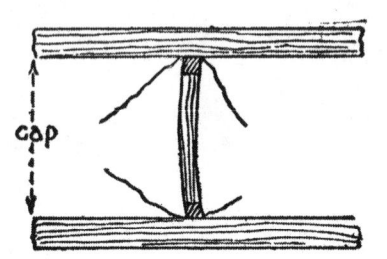

Strut bent throwing wires and gap out of adjustment.

The only exception to the above condition is what is known as the Arch. For instance, in the case of the Maurice Farman, the spars of the centre section plane, which have to take the weight of the nacelle, are arched upwards. If this was not done, it is possible that rough landings might result in the weight causing the spars to become slightly distorted downwards. That would produce a dangerous bending stress, but, as long as the wood is arched, or, at any rate, kept from bending downwards, it will remain in direct compression and no danger can result.

2. *Struts and spars must be symmetrical.* By that I mean that the cross-sectional dimensions must be correct, as otherwise there will be bulging places on the outside, with the result that the stress will not be evenly disposed about the centre of strength, and a bending stress may be produced.

3. *Struts, spars, etc., must be undamaged.* Remember

that, from what I have already explained about bending stresses, the outside fibres of the wood are doing by far the most work. If these get bruised or scored, then the strut or spar suffers in strength much more than one might think at first sight; and, if it ever gets a tendency to bend, it is likely to collapse at that point.

4. *The wood must have a good, clear grain with no cross-grain, knots, or shakes.* Such blemishes produce weak places and, if a tendency to bend appears, then it may collapse at such a point.

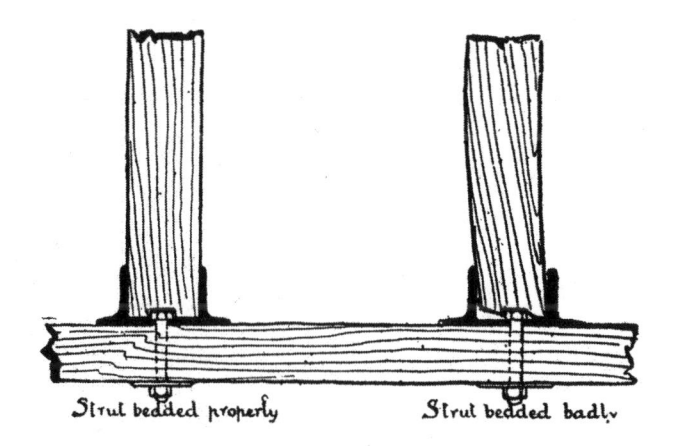

Strut bedded properly   *Strut bedded badly*

5. *The struts, spars, etc., must be properly bedded into their sockets or fittings.* To begin with, they must be of good pushing or gentle tapping fit. They must never be driven in with a heavy hammer. Then again, a strut must bed well down all over its cross-sectional area as illustrated above; otherwise the stress of compression will not be evenly disposed about the centre of strength, and that may produce a bending stress. The bottom of the strut or spar should be covered with some sort of paint, bedded into the socket or fitting, and then withdrawn to see if the paint has stuck all over the bed.

6. The atmosphere is sometimes much damper than at other times, and this causes wood to expand and contract appreciably. This would not matter but for the fact that it does not expand and contract

uniformly, but becomes unsymmetrical, *i.e.,* distorted. I have already explained the danger of that in condition 2. This should be minimized by *well varnishing the wood* to keep the moisture out of it.

FUNCTION OF INTERPLANE STRUTS.—These struts have to keep the lifting surfaces or "planes" apart, but this is only part of their work. They must keep the planes apart, so that the latter are in their correct attitude. That is only so when the spars of the bottom plane are parallel with those of the top plane. Also, the chord of the top plane must be parallel with the chord of the bottom plane. If that is not so, then one plane will not have the same angle of incidence as the other one. At first sight one might think that all that is necessary is to cut all the struts to be the same length, but that is not the case.

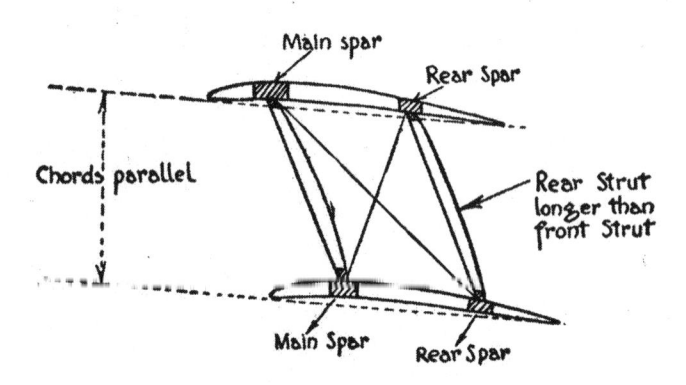

Sometimes, as illustrated above, the rear spar is not so thick as the main spar, and it is then necessary to make up for that difference by making the rear struts correspondingly longer. If that is not done, then the top and bottom chords will not be parallel, and the top and bottom planes will have different angles of incidence. Also, the sockets or fittings, or even the spars upon which they are placed, sometimes vary in thickness owing to faulty manufacture. This must be offset by altering the length of the struts. The best way to proceed is to measure the distance between the top and bottom spars by the side of each strut, and if that distance, or "gap" as it is called, is not as stated in the aeroplane's specifications, then make it correct by

RIGGING

changing the length of the strut. This applies to both front and rear interplane struts. When measuring the gap, always be careful to measure from the centre of the spar, as it may be set at an angle, and the rear of it may be considerably lower than its front.

BORING HOLES IN WOOD.—It should be a strict rule that no spar be used which has an unnecessary hole in it. Before boring a hole, its position should be confirmed by whoever is in charge of the workshop. A bolt-hole should be of a size to enable the bolt to be pushed in, or, at any rate, not more than gently tapped in. Bolts should not be hammered in, as that may split the spar. On the other hand, a bolt should not be slack in its hole, as, in such a case, it may work sideways and split the spar, not to speak of throwing out of adjustment the wires leading from the lug or socket under the bolt-head.

WASHERS.—Under the bolt-head, and also under the nut, a washer must be placed—a very large washer compared with the size which would be used in all-metal construction. This is to disperse the stress over a large area; otherwise the washer may be pulled into the wood and weaken it, besides possibly throwing out of adjustment the wires attached to the bolt or the fitting it is holding to the spar.

LOCKING.—Now as regards locking the bolts. If split pins are used, be sure to see that they are used in such a way that the nut cannot possibly unscrew at all. The split pin should be passed through the bolt as near as possible to the nut. It should not be passed through both nut and bolt.

If it is locked by burring over the edge of the bolt, do not use a heavy hammer and try to spread the whole head of the bolt. That might damage the woodwork inside the fabric-covered surface. Use a small, light hammer, and gently tap round the edge of the bolt until it is burred over.

TURNBUCKLES.—A turnbuckle is composed of a central barrel into each end of which is screwed an eye-bolt. Wires are taken from the eyes of the eye-bolt, and so, by turning the barrel, they can be adjusted to their proper tension. Eye-bolts must be a good fit in the barrel; that is to say, not slack and not very tight. Theoretically it is not necessary to screw the eye-bolt into the barrel for a distance greater than the diameter of the bolt, but, in practice, it is better to screw it in for a considerably greater distance than that if a reasonable degree of safety is to be secured.

Now about turning the barrel to secure the right adjustment. The barrel looks solid, but, as a matter of fact, it is hollow and much more frail than it appears. For that reason it should not be turned by seizing it with pliers, as that may distort it and spoil the bore within it. The best method is to pass a piece of wire through the hole in its centre, and to use that as a lever. When the correct adjustment has been secured, the turnbuckle must be locked to prevent it from unscrewing. It is quite possible to lock it in such a way as to allow it to unscrew a quarter or a half turn, and that would throw the wires out of the very fine adjustment necessary. The proper way is to use the locking wire so that its direction is such as to oppose the tendency of the barrel to unscrew, thus:

WIRES.—The following points should be carefully observed where wire is concerned:

1. *Quality.*—It must not be too hard or too soft. An easy practical way of learning to know the approximate quality of wire is as follows:

Take three pieces, all of the same gauge, and each about a foot in length. One piece should be too soft, another too hard, and the third piece of the right quality. Fix them in a vice, about an inch apart and in a vertical position, and with the light from a window shining upon them. Burnish them if necessary, and you will see a band of light reflected from each wire.

Now bend the wires over as far as possible and away from the light. Where the soft wire is concerned, it will squash out at the bend, and this will be indicated by the band of light, which will broaden at that point. In the case of the wire which is too hard, the band of light will broaden very

RIGGING

little at the turn, but, if you look carefully, you will see some little roughnesses of surface. In the case of the wire of the right quality, the band of light may broaden a very little at the turn, but there will be no roughnesses of surface.

By making this experiment two or three times one can soon learn to know really bad wire from good, and also learn to know the strength of hand necessary to bend the right quality.

2. It must not be damaged. That is to say, it must be unkinked, rustless, and unscored.

3. Now as regards keeping wire in good condition. Where outside wires are concerned, they should be kept *well greased or oiled,* especially where bent over at the ends. Internal bracing wires cannot be reached for the purpose of regreasing them, as they are inside fabric-covered surfaces. They should be prevented from rusting by being painted with an anti-rust mixture. Great care should be taken to see that the wire is perfectly clean and dry before being painted. A greasy finger-mark is sufficient to stop the paint from sticking to the wire. In such a case there will be a little space between the paint and the wire. Air may enter there and cause the wire to rust.

4. *Tension of Wires.*—The tension to which the wires are adjusted is of the greatest importance. All the wires should be of the same tension when the aeroplane is supported in such a way as to throw no stress upon them. If some wires are in greater tension than others, the aeroplane will quickly become distorted and lose its efficiency.

In order to secure the same tension of all wires, the aeroplane, when being rigged, should be supported by packing underneath the lower surfaces as well as by packing underneath the fuselage or nacelle. In this way the anti-lift wires are relieved of the weight, and there is no stress upon any of the wires.

As a general rule the wires of an aeroplane are tensioned too much. The tension should be sufficient to keep the framework rigid. Anything more than that lowers the factor of safety, throws various parts of the framework into undue compression, pulls the fittings into the wood, and will, in the end, distort the whole framework of the aeroplane.

Only experience will teach the rigger what tension to employ. Much may be done by learning the construction of the various types of aeroplanes, the work the various parts do, and in cultivating a touch for tensioning wires by constantly handling them.

5. *Wires with no Opposition Wires.*—In some few cases wires will be found which have no opposition wires pulling in the opposite direction. For instance, an auxiliary lift wire may run from the bottom of a strut to a spar in the top plane at a point between struts. In such a case great care should be taken not to tighten the wire beyond barely taking up the slack.

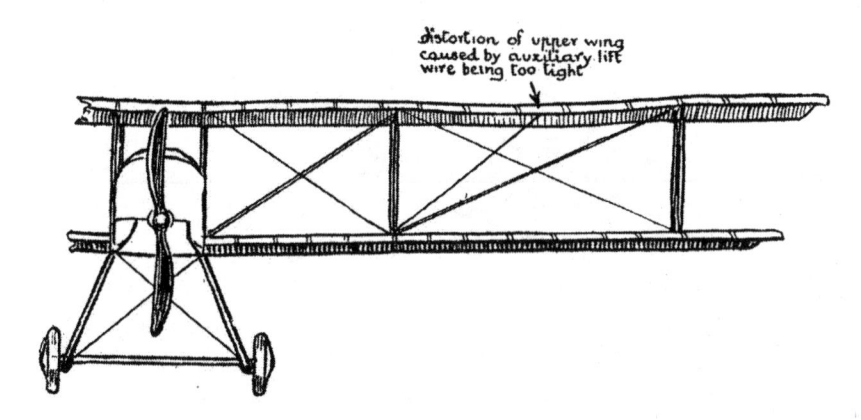

Such a wire must be a little slack, or, as illustrated above, it will distort the framework. That, in the example given, will spoil the camber (curvature) of the surface, and result in changing both the lift and the drift at that part of the surface. Such a condition will cause the aeroplane to lose its directional stability and also to fly one wing down.

I cannot impress this matter of tension upon the reader too strongly. It is of the utmost importance. When this, and also accuracy in securing the various adjustments, has been learned, one is on the way to becoming a good rigger.

6. *Wire Loops.*—Wire is often bent over at its end in the form of a loop, in order to connect with a turnbuckle or fitting. These loops, even when made as perfectly as possible, have a tendency to elongate, thus spoiling the adjustment of the wires. Great care should be taken to minimize this

RIGGING

as far as possible. The rules to be observed are as follows:

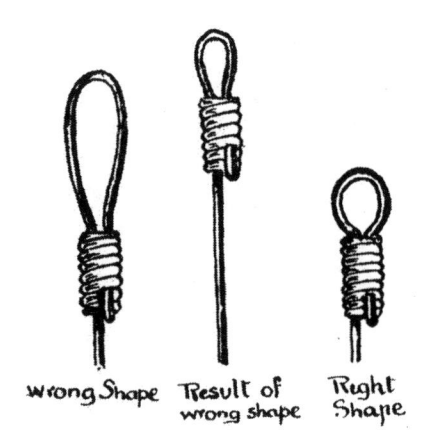

wrong Shape   *Result of wrong shape*   *Right Shape*

(a) The size of the loop should be as small as possible within reason. By that I mean it should not be so small as to create the possibility of the wire breaking.
(b) The shape of the loop should be symmetrical.
(c) It should have well-defined shoulders in order to prevent the ferrule from slipping up. At the same time, a shoulder should not have an angular place.
(d) When the loop is finished it should be undamaged, and it should not be, as is often the case, badly scored.

7. Stranded Wire Cable.—No splice should be served with twine until it has been inspected by whoever is in charge of the workshop. The serving may cover bad work.

Should a strand become broken, then the cable should be replaced at once by another one.

Control cables have a way of wearing out and fraying wherever they pass round pulleys. Every time an aeroplane comes down from flight the rigger should carefully examine the cables, especially where they pass round pulleys. If he finds a strand broken, he should replace the cable.

The ailerons' balance cable on the top of the top plane is often forgotten, since it is necessary to fetch a high pair of steps in order to examine it. Don't slack this, or some gusty day the pilot may unexpectedly find himself minus the aileron control.

THE AEROPLANE SPEAKS

CONTROLLING SURFACES.—The greatest care should be exercised in rigging the aileron, rudder, and elevator properly, for the pilot entirely depends upon them in managing the aeroplane.

Non lifting surface

Position in which controlling surface must be rigged. It will be its position during flight

The ailerons and elevator should be rigged so that, when the aeroplane is in flight, they are in a fair true line with the surface in front and to which they are hinged.

Lifting Surface

Position during flight

Position in which controlling surface must be rigged

If the surface to which they are hinged is not a lifting surface, then they should be rigged to be in a fair true line with it as illustrated above.

If the controlling surface is, as illustrated, hinged to the back of a lifting surface, then it should be rigged a little below the position it would occupy if in a fair true line with the surface in front. This is because, in such a case, it is set at an angle of incidence. This angle will, during flight, cause it to lift a little above the position in which it has been rigged. It is able to lift owing to a certain amount of slack in the control wire holding it—and one cannot adjust the control wire to have no slack, because that would cause it to bind against the pulleys and make the operation of it too hard for the pilot. It is therefore necessary to rig it a little below the position it would occupy if it was rigged in a fair true line with the surface in front. Remember that this only applies when it is hinged to a lifting surface. The greater the angle of incidence (and therefore the lift) of the surface in front, then the more the controlling surface will have to be rigged down.

As a general rule it is safe to rig it down so that its trailing

RIGGING

edge is ½ to ¾ inch below the position it would occupy if in a fair line with the surface in front; or about ½ inch down for every 18 inches of chord of the controlling surface.

When making these adjustments the pilot's control levers should be in their neutral positions. It is not sufficient to lash them. They should be rigidly blocked into position with wood packing.

The surfaces must not be distorted in any way. If they are held true by bracing wires, then such wires must be carefully adjusted. If they are distorted and there are no bracing wires with which to true them, then some of the internal framework will probably have to be replaced.

The controlling surfaces should never be adjusted with a view to altering the stability of the aeroplane. Nothing can be accomplished in that way. The only result will be to spoil the control of the aeroplane.

FABRIC-COVERED SURFACES.—First of all make sure that there is no distortion of spars or ribs, and that they are perfectly sound. Then adjust the internal bracing wires so that the ribs are parallel to the direction of flight. The ribs usually cause the fabric to make a ridge where they occur, and, if such ridge is not parallel to the direction of flight, it will produce excessive drift. As a rule the ribs are at right angles to both main and rear spars.

The tension of the internal bracing wires should be just sufficient to give rigidity to the framework. They should not be tensioned above that unless the wires are, at their ends, bent to form loops. In that case a little extra tension may be given to offset the probable elongation of the loops.

The turnbuckles must now be generously greased, and served round with adhesive tape. The wires must be rendered perfectly dry and clean, and then painted with an anti-rust mixture. The woodwork must be well varnished.

If it is necessary to bore holes in the spars for the purpose of receiving, for instance, socket bolts, then their places should be marked before being bored and their positions confirmed by whoever is in charge of the workshop. All is now ready for the sail-maker to cover the surface with fabric.

ADJUSTMENT OF CONTROL CABLES.—The adjustment of the control cables is quite an art, and upon it will depend to a large degree the quick and easy control of the aeroplane by the pilot.

The method is as follows:

After having rigged the controlling surfaces, and as far as possible secured the correct adjustment of the control cables, then remove the packing which has kept the control levers rigid. Then, sitting in the pilot's seat, move the control levers *smartly*. Tension the control cables so that when the levers are smartly moved there is no perceptible snatch or lag. Be careful not to tension the cables more than necessary to take out the snatch. If tensioned too much they will (1) bind round the pulleys and result in hard work for the pilot; (2) throw dangerous stresses upon the controlling surfaces, which are of rather flimsy construction; and (3) cause the cables to fray round the pulleys quicker than would otherwise be the case.

Now, after having tensioned the cables sufficiently to take out the snatch, place the levers in their neutral positions, and move them to and fro about $\frac{1}{8}$ inch either side of such positions. If the adjustment is correct, it should be possible to see the controlling surfaces move. If they do not move, then the control cables are too slack.

FLYING POSITION.—Before rigging an aeroplane or making any adjustments it is necessary to place it in what is known as its "flying position." I may add that it would be better termed its "rigging position."

In the case of an aeroplane fitted with a stationary engine this is secured by packing up the machine so that the engine foundations are perfectly horizontal both longitudinally and laterally. This position is found by placing a straight-edge and a spirit-level across the engine foundations (both longitudinally and laterally), and great care should be taken to see that the bubble is exactly in the centre of the level. The slightest error will assume magnitude towards the extremities of the aeroplane. Great care should be taken to block up the aeroplane rigidly. In case it gets accidentally disturbed while the work is going on, it is well to constantly verify the flying position by running the straight-edge and spirit-level

RIGGING

over the engine foundations. The straight-edge should be carefully tested before being used, as, being generally made of wood, it will not remain true long. Place it lightly in a vice, and in such a position that a spirit-level on top shows the bubble exactly in the centre. Now slowly move the level along the straight-edge, and the bubble should remain exactly in the centre. If it does not do so, then the straight-edge is not true and must be corrected. *This should never be omitted.*

In the case of aeroplanes fitted with engines of the rotary type, the "flying position" is some special attitude laid down in the aeroplane's specifications, and great care should be taken to secure accuracy.

ANGLE OF INCIDENCE.—One method of finding the angle of incidence is as follows:

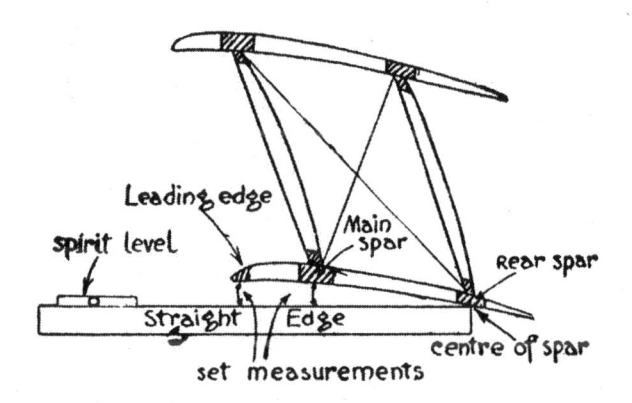

First place the aeroplane in its flying position. The corner of the straight-edge must be placed underneath and against the *centre* of the rear spar, and held in a horizontal position parallel to the ribs. This is secured by using a spirit-level. The set measurement will then be from the top of the straight-edge to the centre of the bottom surface of the main spar, or it may be from the top of the straight-edge to the lowest part of the leading edge. Care should be taken to measure from the centre of the spar and to see that the bubble is exactly in the centre of the level. Remember that all this will be useless if the aeroplane has not been placed accurately in its flying position.

This method of finding the angle of incidence must be

used under every part of the lower surface where struts occur. It should not be used between the struts, because, in such places, the spars may have taken a slight permanent set up or down; not, perhaps, sufficiently bad to make any material difference to the flying of the machine, but quite bad enough to throw out the angle of incidence, which cannot be corrected at such a place.

If the angle is wrong, it should then be corrected as follows:

If it is too great, then the rear spar must be warped up until it is right, and this is done by slackening *all* the wires going to the top of the strut, and then tightening *all* the wires going to the bottom of the strut.

If the angle is too small, then slacken *all* the wires going to the bottom of the strut, and tighten *all* the wires going to the top of the strut, until the correct adjustment is secured.

Never attempt to adjust the angle by warping the main spar.

The set measurement, which is of course stated in the aeroplane's specifications, should be accurate to $\frac{1}{16}$ inch.

LATERAL DIHEDRAL ANGLE.—One method of securing this is as follows, and this method will, at the same time, secure the correct angle of incidence:

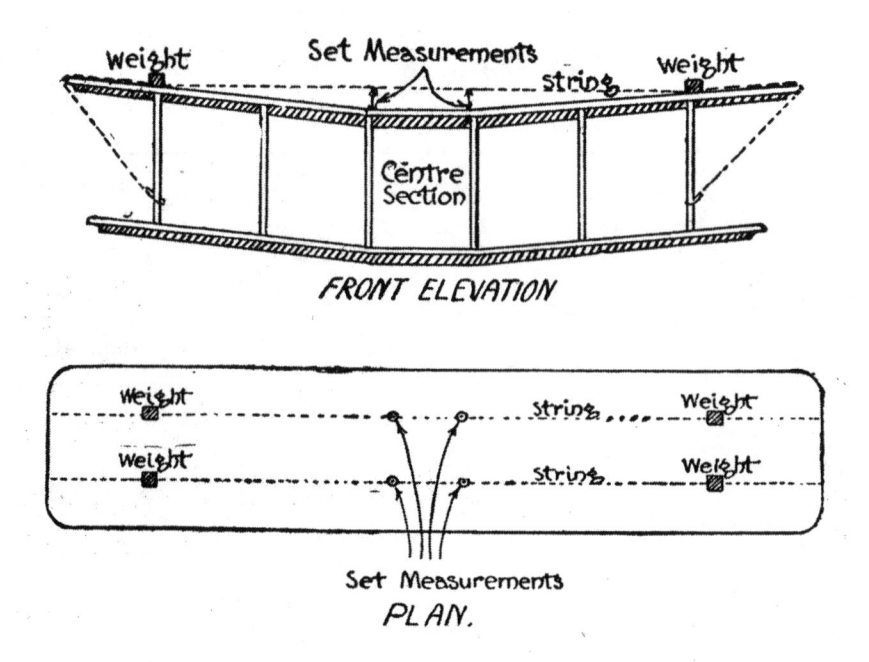

FRONT ELEVATION

PLAN.

RIGGING

The strings, drawn very tight, must be taken over both the main and rear spars of the top surface. They must run between points on the spars just inside the outer struts. The set measurement (which should be accurate to $\frac{1}{16}$ inch or less) is then from the strings down to four points on the main and rear spars of the centre-section surface. These points should be just inside the four centre-section struts; that is to say, as far as possible away from the centre of the centre-section. Do not attempt to take the set measurement near the centre of the centre-section.

The strings should be as tight as possible, and, if it can be arranged, the best way to accomplish that is as shown in the above illustration, *i.e.*, by weighting the strings down to the spars by means of weights and tying each end of the strings to a strut. This will give a tight and motionless string.

However carefully the above adjustment is made, there is sure to be some slight error. This is of no great importance, provided it is divided equally between the left- and right-hand wings. In order to make sure of this, certain check measurements should be taken as follows:

Each bay must be diagonally measured, and such measurements must be the same to within $\frac{1}{16}$ inch on each side of the aeroplane. As a rule such diagonal measurements are taken from the bottom socket of one strut to the top socket of another strut, but this is bad practice, because of possible inaccuracies due to faulty manufacture.

The points between which the diagonal measurements are taken should be at fixed distances from the butts of the spars, such distances being the same on each side of the aeroplane, thus:

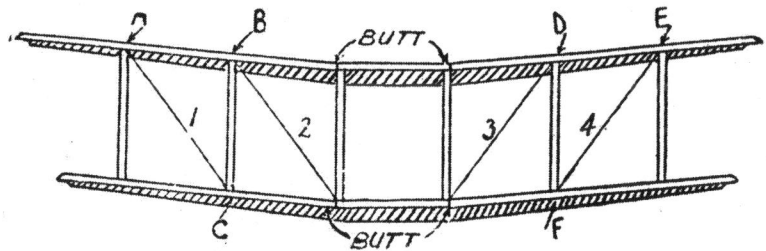

Points A, B, and C, must be the same fixed distances from the butt as are Points D, E, and F. Distances 1 and 2 must equal distances 3 and 4.

The above applies to both front and rear bays.

THE AEROPLANE SPEAKS

It would be better to use the centre line of the aeroplane rather than the butts of the spars. It is not practicable to do so, however, as the centre line probably runs through the petrol tanks, etc.

THE DIHEDRAL BOARD.—Another method of securing the dihedral angle, and also the angle of incidence, is by means of the dihedral board. It is a light handy thing to use, but leads to many errors, and should not be used unless necessary. The reasons are as follows:

The dihedral board is probably not true. If it must be used, then it should be very carefully tested for truth beforehand. Another reason against its use is that it has to be placed on the spars in a position between the struts, and that is just where the spars may have a little permanent set up or down, or some inaccuracy of surface which will, of course, throw out the accuracy of the adjustment. The method of using it is as follows:

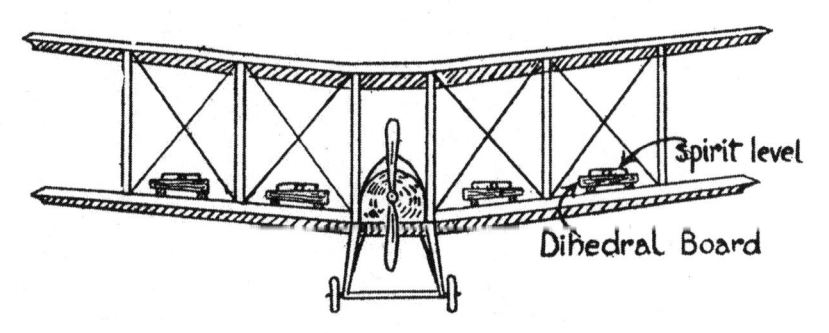

The board is cut to the same angle as that specified for the upward inclination of the surface towards its wing-tips. It is placed on the spar as indicated above, and it is provided with two short legs to raise it above the flanges of the ribs (which cross over the spars), as they may vary in depth. A spirit-level is then placed on the board, and the wires must be adjusted to give the surface such an inclination as to result in the bubble being in the centre of the level. This operation must be performed in respect of each bay both front and rear. The bays must then be diagonally measured as already explained.

YET ANOTHER METHOD of finding the dihedral angle, and at the same time the angle of incidence, is as follows:

RIGGING

A horizontal line is taken from underneath the butt of each spar, and the set measurement is either the angle it makes with the spar, or a fixed measurement from the line to the spar taken at a specified distance from the butt. This operation must be performed in respect of both main and rear spars, and all the bays must be measured diagonally afterwards.

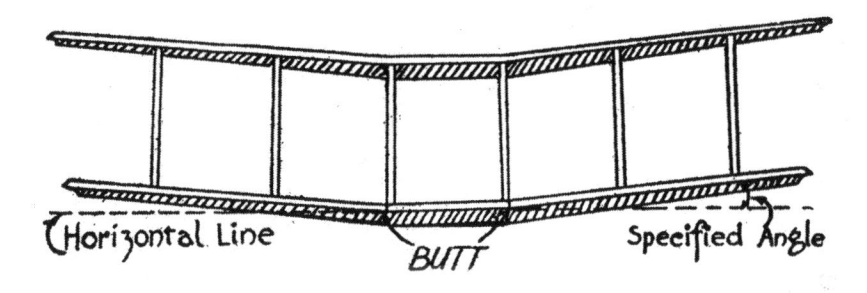

Whichever method is used, be sure that after the job is done the spars are perfectly straight.

STAGGER.—The stagger is the distance the top surface is in advance of the bottom surface when the aeroplane is in flying position. The set measurement is obtained as follows:

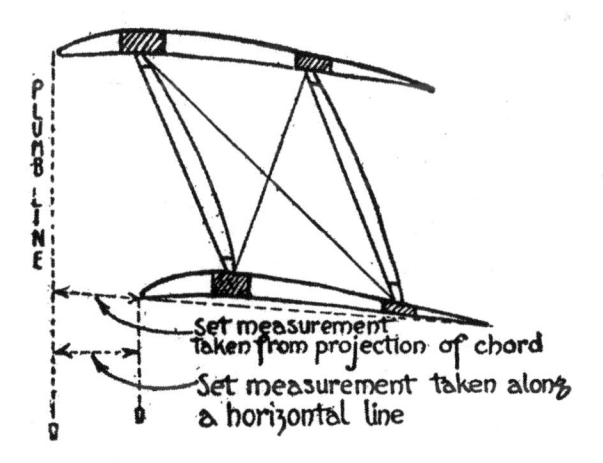

Plumb-lines must be dropped over the leading edge of the top surface wherever struts occur, and also near the fuselage. The set measurement is taken from the front of the lower leading edge to the plumb-lines. It makes a difference whether the measurement is taken along a horizontal line

THE AEROPLANE SPEAKS

(which can be found by using a straight-edge and a spirit-level) or along a projection of the chord. The line along which the measurement should be taken is laid down in the aeroplane's specifications.

If a mistake is made and the measurement taken along the wrong line, it may result in a difference of perhaps ¼ inch or more to the stagger, with the certain result that the aeroplane will, in flight, be nose-heavy or tail-heavy.

After the adjustments of the angles of incidence, dihedral, and stagger have been secured, it is as well to confirm all of them, as, in making the last adjustment, the first one may have been spoiled.

OVER-ALL ADJUSTMENTS.—The following over-all check measurements should now be taken.

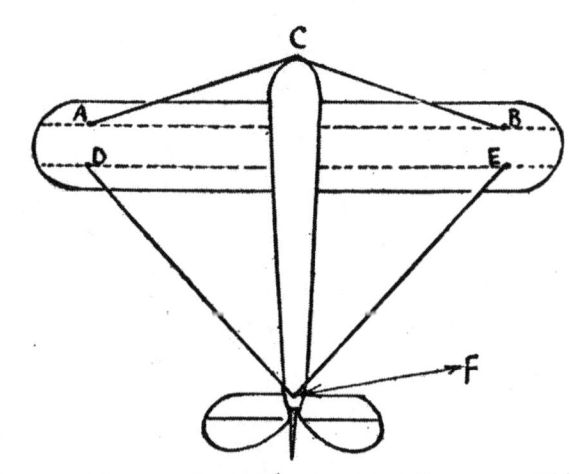

The dotted lines on the surface represent the spars within it.

The straight lines AC and BC should be equal to within ⅛ inch. The point C is the centre of the propeller, or, in the case of a "pusher" aeroplane, the centre of the nacelle. The points A and B are marked on the main spar, and must in each case be the same distance from the butt of the spar. The rigger should not attempt to make A and B merely the sockets of the outer struts, as they may not have been placed quite accurately by the manufacturer. The lines AC and BC must be taken from both top and bottom spars—two measurements on each side of the aeroplane.

The two measurements FD and FE should be equal to

RIGGING

within ⅛ inch. F is the centre of the fuselage or rudder-post. D and E are points marked on both top and bottom rear spars, and each must be the same fixed distance from the butt of the spar. Two measurements on each side of the aeroplane.

If these over-all measurements are not correct, then it is probably due to some of the drift or anti-drift wires being too tight or too slack. It may possibly be due to the fuselage being out of truth, but of course the rigger should have made quite sure that the fuselage was true before rigging the rest of the machine. Again, it may be due to the internal bracing wires within the lifting surfaces not being accurately adjusted, but of course this should have been seen to before covering the surfaces with fabric.

FUSELAGE.—The method of truing the fuselage is laid down in the aeroplane's specifications. After it has been adjusted according to the specified directions, it should then be arranged on trestles in such a way as to make about three-quarters of it towards the tail stick out unsupported. In this way it will assume a condition as near as possible to flying conditions, and when it is in this position the set measurements should be confirmed. If this is not done it may be out of truth, but perhaps appear all right when supported by trestles at both ends, as, in such case, its weight may keep it true as long as it is resting upon the trestles.

THE TAIL-PLANE (EMPENNAGE).—The exact angle of incidence of the tail-plane is laid down in the aeroplane's specifications. It is necessary to make sure that the spars are horizontal when the aeroplane is in flying position and the tail unsupported as explained above under the heading of Fuselage. If the spars are tapered, then make sure that their centre lines are horizontal.

UNDERCARRIAGE.—The undercarriage must be very carefully aligned as laid down in the specifications.

1. The aeroplane must be placed in its flying position and sufficiently high to ensure the wheels being off the ground when rigged. When in this position the axle must be hori-

zontal and the bracing wires adjusted to secure the various set measurements stated in the specifications.

2. Make sure that the struts bed well down into their sockets.

3. Make sure that the shock absorbers are of equal tension. In the case of rubber shock absorbers, both the number of turns and the lengths must be equal.

How to Diagnose Faults in Flight, Stability, and Control.

Directional Stability will be badly affected if there is more drift (*i.e.*, resistance) on one side of the aeroplane than there is on the other side. The aeroplane will tend to turn towards the side having the most drift. This may be caused as follows:

1. The angle of incidence of the main surface or the tail surface may be wrong. The greater the angle of incidence, the greater the drift. The less the angle, the less the drift.

2. If the alignment of the fuselage, fin in front of the rudder, the struts or stream-line wires, or, in the case of the Maurice Farman, the front outriggers, are not absolutely correct—that is to say, if they are turned a little to the left or to the right instead of being in line with the direction of flight—then they will act as a rudder and cause the aeroplane to turn off its course.

3. If any part of the surface is distorted, it will cause the aeroplane to turn off its course. The surface is cambered, *i.e.*, curved, to pass through the air with the least possible drift. If, owing perhaps to the leading edge, spars, or trailing edge becoming bent, the curvature is spoiled, that will result in changing the amount of drift on one side of the aeroplane, which will then have a tendency to turn off its course.

Lateral Instability (Flying One Wing Down).—The only possible reason for such a condition is a difference in the lifts of right and left wings. That may be caused as follows:

1. The angle of incidence may be wrong. If it is too great, it will produce more lift than on the other side of the aeroplane; and if too small, it will produce less lift than on

RIGGING

the other side—the result being that, in either case, the aeroplane will try to fly one wing down.

2. *Distorted Surfaces.*—If some part of the surface is distorted, then its camber is spoiled, and the lift will not be the same on both sides of the aeroplane, and that, of course, will cause it to fly one wing down.

LONGITUDINAL INSTABILITY may be due to the following reasons:

1. *The stagger may be wrong.* The top surface may have drifted back a little owing to some of the wires, probably the incidence wires, having elongated their loops or having pulled the fittings into the wood. If the top surface is not staggered forward to the correct degree, then consequently the whole of its lift is too far back, and it will then have a tendency to lift up the tail of the machine too much. The aeroplane would then be said to be "nose-heavy."

A $\frac{1}{4}$-inch area in the stagger will make a very considerable difference to the longitudinal stability.

2. If *the angle of incidence* of the main surface is not right, it will have a bad effect, especially in the case of an aeroplane with a lifting tail-plane.

If the angle is too great, it will produce an excess of lift, and that may lift up the nose of the aeroplane and result in a tendency to fly "tail-down." If the angle is too small, it will produce a decreased lift, and the aeroplane may have a tendency to fly "nose-down."

3. *The fuselage* may have become warped upward or downward, thus giving the tail-plane an incorrect angle of incidence. If it has too much angle, it will lift too much, and the aeroplane will be "nose-heavy." If it has too little angle, then it will not lift enough, and the aeroplane will be "tail-heavy."

4. (The least likely reason.) *The tail-plane* may be mounted upon the fuselage at a wrong angle of incidence, in which case it must be corrected. If nose-heavy, it should be given a smaller angle of incidence. If tail-heavy, it should be given a larger angle; but care should be taken not to give it too great an angle, because the longitudinal stability entirely depends upon the tail-plane being set at a much

smaller angle of incidence than is the main surface, and if that difference is decreased too much, the aeroplane will become uncontrollable longitudinally. Sometimes the tail-plane is mounted on the aeroplane at the same angle as the main surface, but it actually engages the air at a lesser angle, owing to the air being deflected downwards by the main surface. There is then, in effect, a longitudinal dihedral as explained and illustrated in Chapter I.

CLIMBS BADLY.—Such a condition is, apart from engine or propeller trouble, probably due to (1) distorted surfaces, or (2) too small an angle of incidence.

FLIGHT SPEED POOR.—Such a condition is, apart from engine or propeller trouble, probably due to (1) distorted surfaces, (2) too great an angle of incidence, or (3) dirt or mud, and consequently excessive skin-friction.

INEFFICIENT CONTROL is probably due to (1) wrong setting of control surfaces, (2) distortion of control surfaces, or (3) control cables being badly tensioned.

WILL NOT "TAXI" STRAIGHT.—If the aeroplane is uncontrollable on the ground, it is probably due to (1) alignment of undercarriage being wrong, or (2) unequal tension of shock absorbers.

CHAPTER IV

THE PROPELLER, OR "AIR-SCREW"

THE sole object of the propeller is to translate the power of the engine into thrust.

The propeller screws through the air, and its blades, being set at an angle inclined to the direction of motion, secure a reaction, as in the case of the aeroplane's lifting surface.

This reaction may be conveniently divided into two component parts or values, namely, Thrust and Drift (see illustration overleaf).

The Thrust is opposed to the Drift of the aeroplane, and must be equal and opposite to it at flying speed. If it falls off in power, then the flying speed must decrease to a velocity, at which the aeroplane drift equals the decreased thrust.

The Drift of the propeller may be conveniently divided into the following component values:

Active Drift, produced by the useful thrusting part of the propeller.

Passive Drift, produced by all the rest of the propeller, *i.e.,* by its detrimental surface.

Skin Friction, produced by the friction of the air with roughnesses of surface.

Eddies attending the movement of the air caused by the action of the propeller.

Cavitation (very marked at excessive speed of revolution). A tendency of the propeller to produce a cavity or semi-vacuum in which it revolves, the thrust decreasing with increase of speed and cavitation.

THRUST-DRIFT RATIO.—The proportion of thrust to drift is of paramount importance, for it expresses the efficiency of the propeller. It is affected by the following factors:

Speed of Revolution.—The greater the speed, the greater the proportion of drift to thrust. This is due to the increase with speed of the passive drift, which carries with it no increase in thrust. For this reason propellers are often geared down to revolve at a lower speed than that of the engine.

Angle of Incidence.—The same reasons as in the case of the aeroplane surface.

Surface Area.—Ditto.

Aspect Ratio.—Ditto.

Camber.—Ditto.

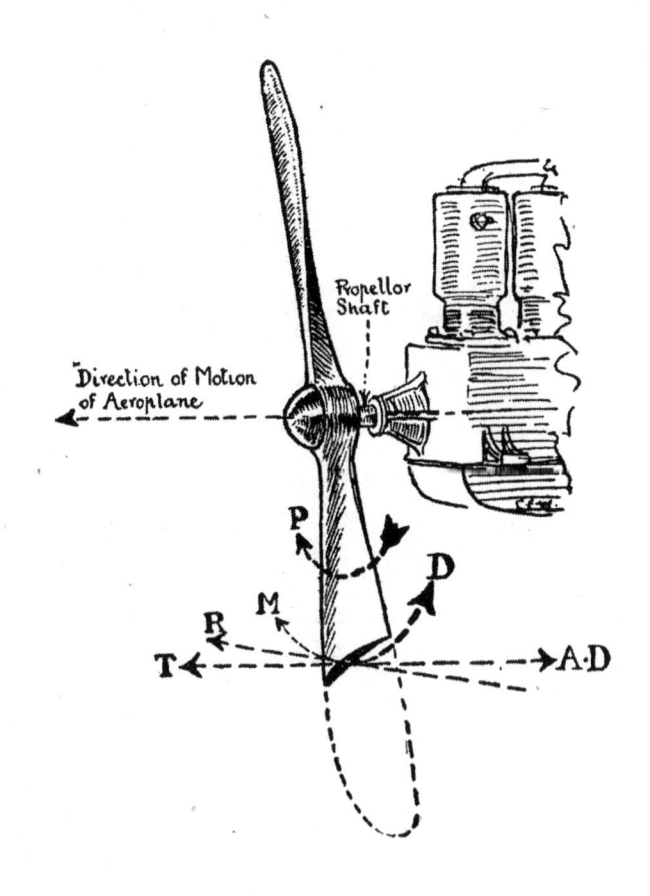

S, Section through propeller blade.
M, Direction of motion of propeller (rotary).
R, Direction of reaction.
T, Direction of thrust.
AD, Direction of the resistance of the air to the passage of the aeroplane, *i.e.*, aeroplane drift.
D, Direction of propeller drift (rotary).
P, Engine power, opposed to propeller drift and transmitted to the propeller through the propeller shaft.

THE PROPELLER, OR "AIR-SCREW"

In addition to the above factors there are, when it comes to actually designing a propeller, mechanical difficulties to consider. For instance, the blades must be of a certain strength and consequent thickness. That, in itself, limits the aspect ratio, for it will necessitate a chord long enough in proportion to the thickness to make a good camber possible. Again, the diameter of the propeller must be limited, having regard to the fact that greater diameters than those used to-day would not only result in excessive weight of construction, but would also necessitate a very high undercarriage to keep the propeller off the ground, and such undercarriage would not only produce excessive drift, but would also tend to make the aeroplane stand on its nose when alighting. The latter difficulty cannot be overcome by mounting the propeller higher, as the centre of its thrust must be approximately coincident with the centre of aeroplane drift.

Maintenance of Efficiency.

The following conditions must be observed:

1. PITCH ANGLE.—The angle, at any given point on the propeller, at which the blade is set is known as the pitch angle, and it must be correct to half a degree if reasonable efficiency is to be maintained.

This angle secures the "pitch," which is the distance the propeller advances during one revolution, supposing the air to be solid. The air, as a matter of fact, gives back to the thrust of the blades just as the pebbles slip back as one ascends a shingle beach. Such "give-back" is known as *Slip*. If a propeller has a pitch of, say, 10 feet, but actually advances, say, only 8 feet owing to slip, then it will be said to possess 20 per cent. slip.

Thus, the pitch must equal the flying speed of the aeroplane plus the slip of the propeller. For example, let us find the pitch of a propeller, given the following conditions:

Flying speed .... 70 miles per hour.
Propeller revolutions .... 1,200 per minute.
Slip .... 15 per cent.

THE AEROPLANE SPEAKS

First find the distance in feet the aeroplane will travel forward in one minute. That is—

$$\frac{369{,}600 \text{ feet (70 miles)}}{60 \text{ " (minutes)}} = 6{,}160 \text{ feet per minute.}$$

Now divide the feet per minute by the propeller revolutions per minute, add 15 per cent. for the slip, and the result will be the propeller pitch:

$$\frac{6{,}160}{1{,}200} + 15 \text{ per cent.} = 5 \text{ feet } 1\tfrac{3}{5} \text{ inches.}$$

In order to secure a constant pitch from root to tip of blade, the pitch angle decreases towards the tip. This is necessary, since the end of the blade travels faster than its root, and yet must advance forward at the same speed as the rest of the propeller. For example, two men ascending a hill. One prefers to walk fast and the other slowly, but they wish to arrive at the top of the hill simultaneously. Then the fast walker must travel a farther distance than the slow one, and his angle of path (pitch angle) must be smaller than the angle of path taken by the slow walker. Their pitch angles are different, but their pitch (in this case altitude reached in a given time) is the same.

THE PROPELLER, OR "AIR-SCREW"

In order to test the pitch angle, the propeller must be mounted upon a shaft at right angles to a beam the face of which must be perfectly level, thus:

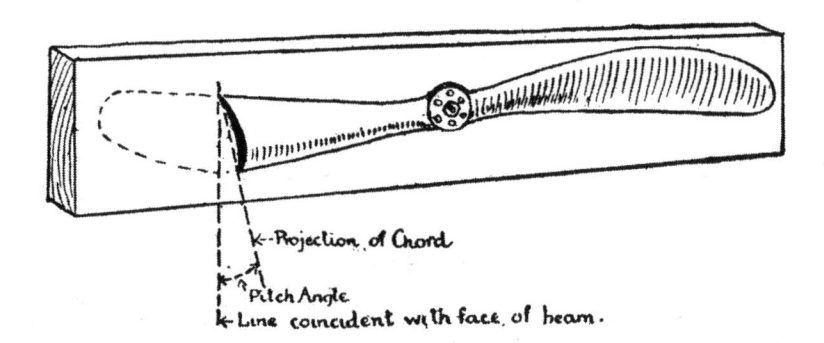

First select a point on the blade at some distance (say about 2 feet) from the centre of the propeller. At that point find, by means of a protractor, the angle a projection of the chord makes with the face of the beam. That angle is the pitch angle of the blade at that point.

Now lay out the angle on paper, thus:

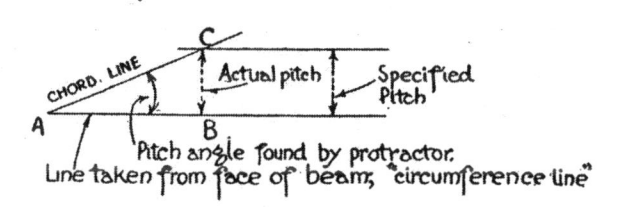

The line above and parallel to the circumference line must be placed in a position making the distance between the two lines equal to the specified pitch, which is, or should be, marked upon the boss of the propeller.

Now find the circumference of the propeller where the pitch angle is being tested. For example, if that place is 2 feet radius from the centre, then the circumference will be 2 feet × 2 = 4 feet diameter, which, if multiplied by 3.1416 = 15.56 feet circumference.

Now mark off the circumference distance, which is represented above by A-B, and reduce it in scale for convenience.

The distance a vertical line makes between B and the chord line is the pitch at the point where the angle is being tested, and it should coincide with the specified pitch. You will note, from the above illustration, that the actual pitch line should meet the junction of the chord line and top line.

The propeller should be tested at several points, about a foot apart, on each blade; and the diagram, provided the propeller is not faulty, will then look like this:

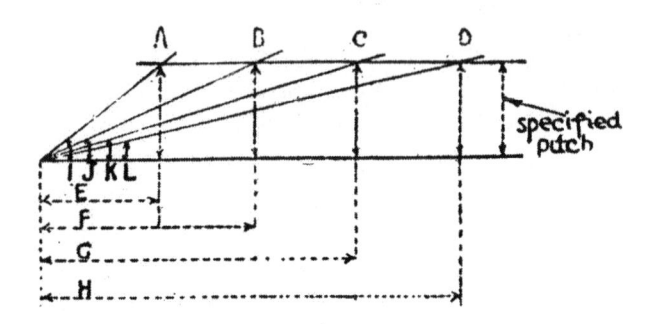

A, B, C, and D, Actual pitch at points tested.
I, Pitch angle at point tested nearest to centre of propeller.
E, Circumference at I.
J, Pitch angle at point tested nearest to I.
F, Circumference at J.
K, Pitch angle at next point tested.
G, Circumference at K.
L, Pitch angle tested at point nearest tip of blade.
H, Circumference at L.

At each point tested the actual pitch coincides with the specified pitch: a satisfactory condition.

A faulty propeller will produce a diagram something like this:

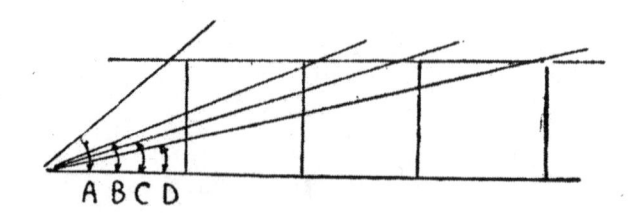

At every point tested the pitch angle is wrong, for nowhere does the actual pitch coincide with the specified pitch. Angles A, C, and D, are too large, and B is too small. The angle should be correct to half a degree if reasonable efficiency is to be maintained.

THE PROPELLER, OR "AIR-SCREW"

A fault in the pitch angle may be due to (1) faulty manufacture, (2) distortion, or (3) the shaft hole through the boss being out of position.

2. STRAIGHTNESS.—To test for straightness the propeller must be mounted upon a shaft. Now bring the tip of one blade round to graze some fixed object. Mark the point it grazes. Now bring the other tip round, and it should come within ⅛ inch of the mark. If it does not do so, it is due to (1) faulty manufacture, (2) distortion, or (3) to the hole through the boss being out of position.

3. LENGTH.—The blades should be of equal length to 1/16 inch.

4. BALANCE.—The usual method of testing a propeller for balance is as follows: Mount it upon a shaft, which must be on ball-bearings. Place the propeller in a horizontal position, and it should remain in that position. If a weight of a trifle over an ounce placed in a bolt-hole on one side of the boss fails to disturb the balance, then the propeller is usually regarded as unfit for use.

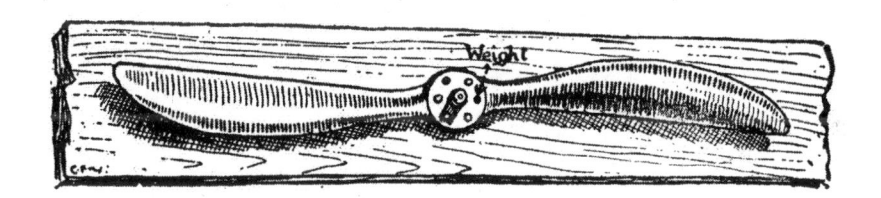

The above method is rather futile, as it does not test for the balance of centrifugal force, which comes into play as soon as the propeller revolves. It can be tested as follows:

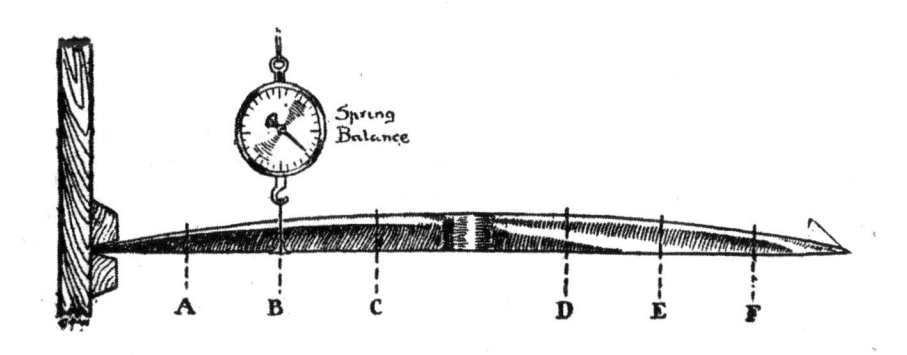

THE AEROPLANE SPEAKS

The propeller must be in a horizontal position, and then weighed at fixed points, such as A, B, C, D, E, and F, and the weights noted. The points A, B, and C must, of course, be at the same fixed distances from the centre of the propeller as the points D, E, and F. Now reverse the propeller and weigh at each point again. Note the results. The first series of weights should correspond to the second series, thus:

Weight A should equal weight F.
" B " " " E.
" C " " " D.

There is no standard practice as to the degree of error permissible, but if there are any appreciable differences the propeller is unfit for use.

5. SURFACE AREA.—The surface area of the blades should be equal. Test with callipers thus:

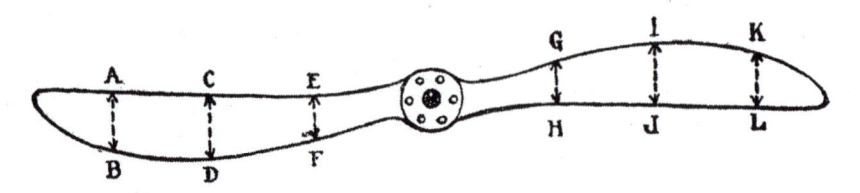

The distance A-B should equal K-L.
,, ,, C-B ,, ,, L-J.
,, ,, E-F ,, ,, G-H.

The points between which the distances are taken must, of course, be at the same distance from the centre in the case of each blade.

There is no standard practice as to the degree of error permissible. If, however, there is an error of over $\frac{1}{8}$ inch, the propeller is really unfit for use.

6. CAMBER.—The camber (curvature) of the blades should be (1) equal, (2) decrease evenly towards the tips of the blades, and (3) the greatest depth of the curve should, at any point of the blade, be approximately at the same percentage of the chord from the leading edge as at other points.

It is difficult to test the top camber without a set of templates, but a fairly accurate idea of the concave camber

THE PROPELLER, OR "AIR-SCREW"

can be secured by slowly passing a straight-edge along the blade, thus:

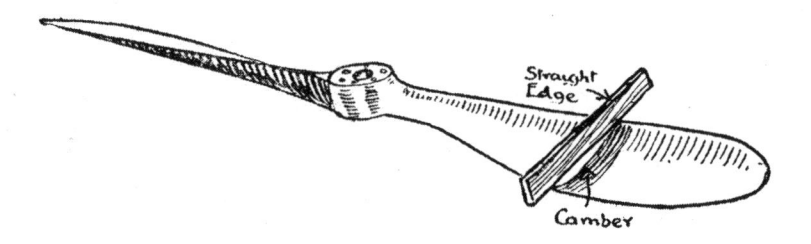

The camber can now be easily seen, and as the straight-edge is passed along the blade, the observer should look for any irregularities of the curvature, which should gradually and evenly decrease towards the tip of the blade.

7. THE JOINTS.—The usual method for testing the glued joints is by revolving the propeller at greater speed than it will be called upon to make during flight, and then carefully examining the joints to see if they have opened. It is not likely, however, that the reader will have the opportunity of making this test. He should, however, examine all the joints very carefully, trying by hand to see if they are quite sound. Suspect a propeller of which the joints appear to hold any thickness of glue. Sometimes the joints in the boss open a little, but this is not dangerous unless they extend to the blades, as the bolts will hold the laminations together.

8. CONDITION OF SURFACE.—The surface should be very smooth, especially towards the tips of the blades. Some propeller tips have a speed of over 30,000 feet a minute, and any roughness will produce a bad drift or resistance and lower the efficiency.

9. MOUNTING.—Great care should be taken to see that the propeller is mounted quite straight on its shaft. Test in the same way as for straightness. If it is not straight, it is possibly due to some of the propeller bolts being too slack or to others having been pulled up too tightly.

FLUTTER.—Propeller "flutter," or vibration, may be due to faulty pitch angle, balance, camber, or surface area. It causes a condition sometimes mistaken for engine trouble, and one which may easily lead to the collapse of the propeller.

THE AEROPLANE SPEAKS

CARE OF PROPELLERS.—The care of propellers is of the greatest importance, as they become distorted very easily.

1. Do not store them in a very damp or a very dry place.
2. Do not store them where the sun will shine upon them.
3. Never leave them long in a horizontal position or leaning up against a wall.
4. They should be hung on horizontal pegs, and the position of the propellers should be vertical.

If the points I have impressed upon you in these notes are not attended to, you may be sure of the following results:

1. Lack of efficiency, resulting in less aeroplane speed and climb than would otherwise be the case.
2. Propeller "flutter" and possible collapse.
3. A bad stress upon the propeller shaft and its bearings.

TRACTOR.—A propeller mounted in front of the main surface.

PUSHER.—A propeller mounted behind the main surface.

FOUR-BLADED PROPELLERS.—Four-bladed propellers are suitable only when the pitch is comparatively large.

For a given pitch, and having regard to "interference," they are not so efficient as two-bladed propellers.

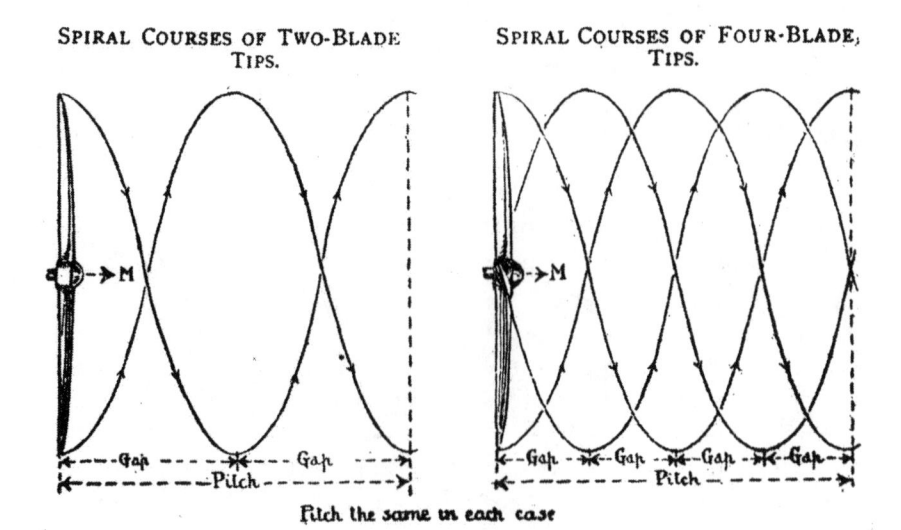

SPIRAL COURSES OF TWO-BLADE TIPS.

SPIRAL COURSES OF FOUR-BLADE TIPS.

Pitch the same in each case

THE PROPELLER, OR "AIR-SCREW"

The smaller the pitch, the less the "gap," *i.e.,* the distance, measured in the direction of the thrust, between the spiral courses of the blades (see illustration on preceding page).

If the gap is too small, then the following blade will engage air which the preceding blade has put into motion, with the result that the following blade will not secure as good a reaction as would otherwise be the case. It is very much the same as in the case of the aeroplane gap.

For a given pitch, the gap of a four-bladed propeller is only half that of a two-bladed one. Therefore the four-bladed propeller is only suitable for large pitch, as such pitch produces spirals with a large gap, thus offsetting the decrease in gap caused by the numerous blades.

The greater the speed of rotation, the less the pitch for a given aeroplane speed. Then, in order to secure a large pitch and consequently a good gap, the four-bladed propeller is usually geared to rotate at a lower speed than would be the case if directly attached to the engine crank-shaft.

CHAPTER V

MAINTENANCE

CLEANLINESS.—The fabric must be kept clean and free from oil, as that will rot it. To take out dirt or oily patches, try acetone. If that will not remedy matters, then try petrol, but use it sparingly, as otherwise it will take off an unnecessary amount of dope. If that will not remove the dirt, then hot water and soap will do so, but, in that case, be sure to use soap having no alkali in it, as otherwise it may injure the fabric. Use the water sparingly, or it may get inside the planes and rust the internal bracing wires, or cause some of the wooden framework to swell.

The wheels of the undercarriage have a way of throwing up mud on to the lower surface. This should, if possible, be taken off while wet. It should never be scraped off when dry, as that may injure the fabric. If dry, then it should be moistened before being removed.

Measures should be taken to prevent dirt from collecting upon any part of the aeroplane, as, otherwise, excessive skin-friction will be produced with resultant loss of flight speed. The wires, being greasy, collect dirt very easily.

CONTROL CABLES.—After every flight the rigger should pass his hand over the control cables and carefully examine them near pulleys. Removal of grease may be necessary to make a close inspection possible. If only one strand is broken the wire should be replaced. Do not forget the aileron balance wire on the top surface.

Once a day try the tension of the control cables by smartly moving the control levers about as explained elsewhere.

WIRES.—All the wires should be kept well greased or oiled, and in the correct tension. When examining the wires, it is necessary to place the aeroplane on level ground, as otherwise it may be twisted, thus throwing some wires into

MAINTENANCE

undue tension and slackening others. The best way, if there is time, is to pack the machine up into its "flying position."

If you see a slack wire, do not jump to the conclusion that it must be tensioned. Perhaps its opposition wire is too tight, in which case slacken it, and possibly you will find that will tighten the slack wire.

Carefully examine all wires and their connections near the propeller, and be sure that they are snaked round with safety wire, so that the latter may keep them out of the way of the propeller if they come adrift.

The wires inside the fuselage should be cleaned and re-greased about once a fortnight.

STRUTS AND SOCKETS.—These should be carefully examined to see if any splitting has occurred.

DISTORTION.—Carefully examine all surfaces, including the controlling surfaces, to see whether any distortion has occurred. If distortion can be corrected by the adjustment of wires, well and good; but if not, then some of the internal framework probably requires replacement.

ADJUSTMENTS.—Verify the angles of incidence, dihedral, and stagger, and the rigging position of the controlling surfaces, as often as possible.

UNDERCARRIAGE.—Constantly examine the alignment and fittings of the undercarriage, and the condition of tyres and shock absorbers. The latter, when made of rubber, wear quickest underneath. Inspect axles and skids to see if there are any signs of them becoming bent. The wheels should be taken off occasionally and greased.

LOCKING ARRANGEMENTS.—Constantly inspect the locking arrangements of turnbuckles, bolts, etc. Pay particular attention to the control cable connections, and to all moving parts in respect of the controls.

LUBRICATION.—Keep all moving parts, such as pulleys, control levers, and hinges of controlling surfaces, well greased.

SPECIAL INSPECTION.—Apart from constantly examining the aeroplane with reference to the above points I have made, I think that, in the case of an aeroplane in constant use it is an excellent thing to make a special inspection of every part, say once a week. This will take from two to three hours, according to the type of aeroplane. In order to carry it out methodically, the rigger should have a list of every part down to the smallest split-pin. He can then check the parts as he examines them, and nothing will be passed over. This, I know from experience, greatly increases the confidence of the pilot, and tends to produce good work in the air.

WINDY WEATHER.—The aeroplane, when on the ground, should face the wind; and it is advisable to lash the control lever fast, so that the controlling surfaces may not be blown about and possibly damaged.

"VETTING" BY EYE.—This should be practised at every opportunity, and, if persevered in, it is possible to become quite expert in diagnosing by eye faults in flight efficiency, stability and control.

The aeroplane should be standing upon level ground, or, better than that, packed up into its "flying position."

Now stand in front of it and line up the leading edge with the main spar, rear spar, and trailing edge. Their shadows can usually be seen through the fabric. Allowance must, of course, be made for wash-in and wash-out; otherwise, the parts I have specified should be parallel with each other.

Now line up the centre part of the main-plane with the tail-plane. The latter should be horizontal.

Next, sight each interplane front strut with its rear strut. They should be parallel.

Then, standing on one side of the aeroplane, sight all the front struts. The one nearest to you should cover all the others. This applies to the rear struts also.

Look for distortion of leading edges, main and rear spars, trailing edges, tail-plane and controlling surfaces.

This sort of thing, if practised constantly, will not only develop an expert eye for diagnosis of faults, but will also greatly assist in impressing upon the memory the characteristics and possible troubles of the various types of aeroplanes.

MISHANDLING OF THE GROUND.—This is the cause of a lot of unnecessary damage. The golden rule to observe is: PRODUCE NO BENDING STRESSES.

MAINTENANCE

Nearly all the wood in an aeroplane is designed to take merely the stress of direct compression, and it cannot be bent safely. Therefore, in packing an aeroplane up from the ground, or in pulling or pushing it about, be careful to stress it in such a way as to produce, as far as possible, only direct compression stresses. For instance, if it is necessary to support the lifting surface, then the packing should be arranged to come directly under the struts so that they may take the stress in the form of compression for which they are designed. Such supports should be covered with soft packing in order to prevent the fabric from becoming damaged.

When pulling an aeroplane along, if possible, pull from the top of the undercarriage struts. If necessary to pull from elsewhere, then do so by grasping the interplane struts as low down as possible.

Never lay fabric-covered parts upon a concrete floor. Any slight movement will cause the fabric to scrape over the floor with resultant damage.

Struts, spars, etc., should never be left about the floor, as in such position they are likely to become scored. I have already explained the importance of protecting the outside fibres of the wood. Remember also that wood becomes distorted easily. This particularly applies to interplane struts. If there are no proper racks to stand them in, then the best plan is to lean them up against the wall in as near a vertical position as possible.

TIME.—Learn to know the time necessary to complete any of the various rigging jobs. This is really important. Ignorance of this will lead to bitter disappointments in civil life; and, where Service flying is concerned, it will, to say the least of it, earn unpopularity with senior officers, and fail to develop respect and good work where men are concerned.

THE AEROPLANE SHED.—This should be kept as clean and orderly as possible. A clean, smart shed produces briskness, energy, and pride of work. A dirty, disorderly shed nearly always produces slackness and poor quality of work, lost tools and mislaid material.

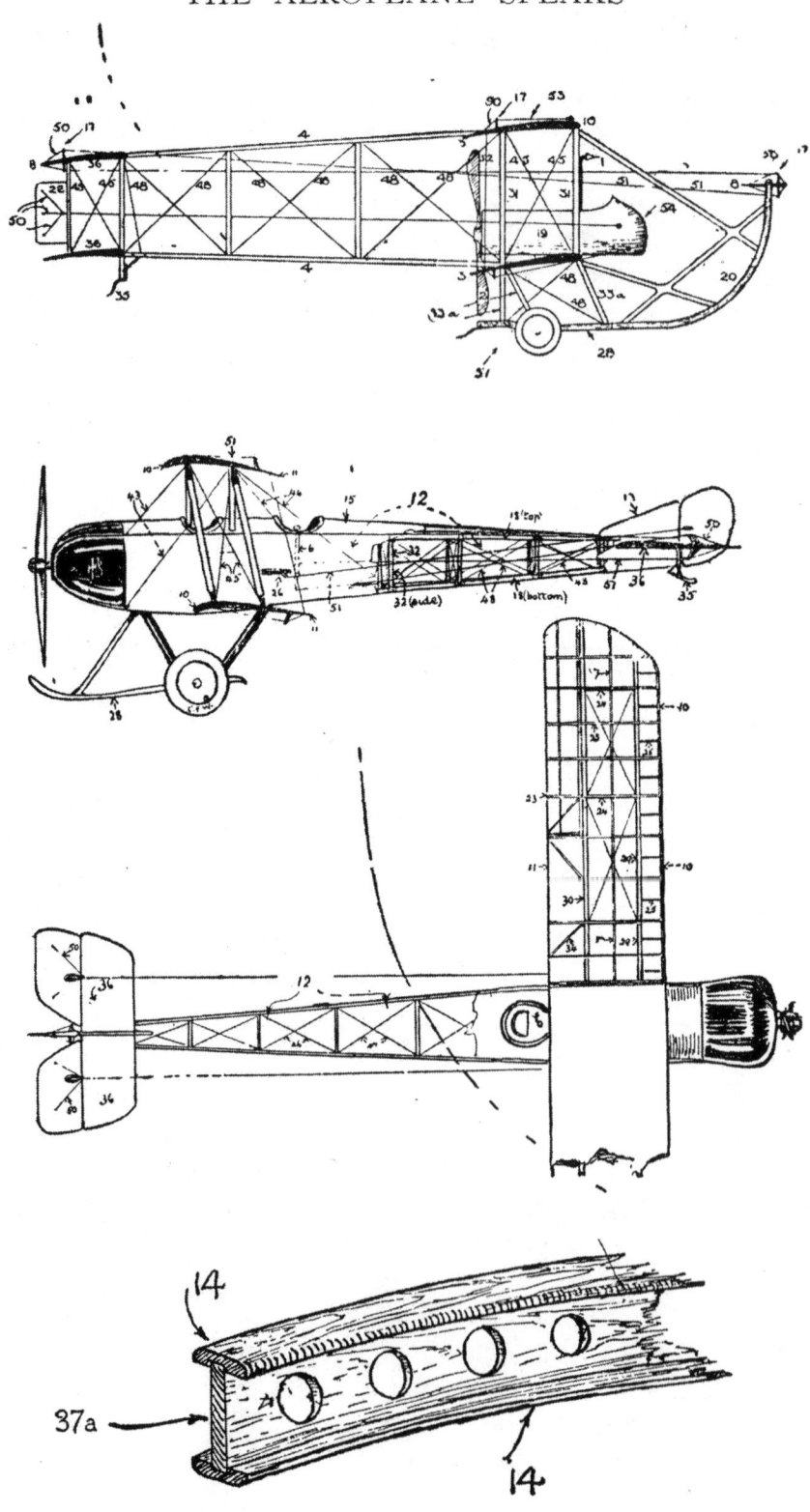

THE AEROPLANE SPEAKS

MAINTENANCE

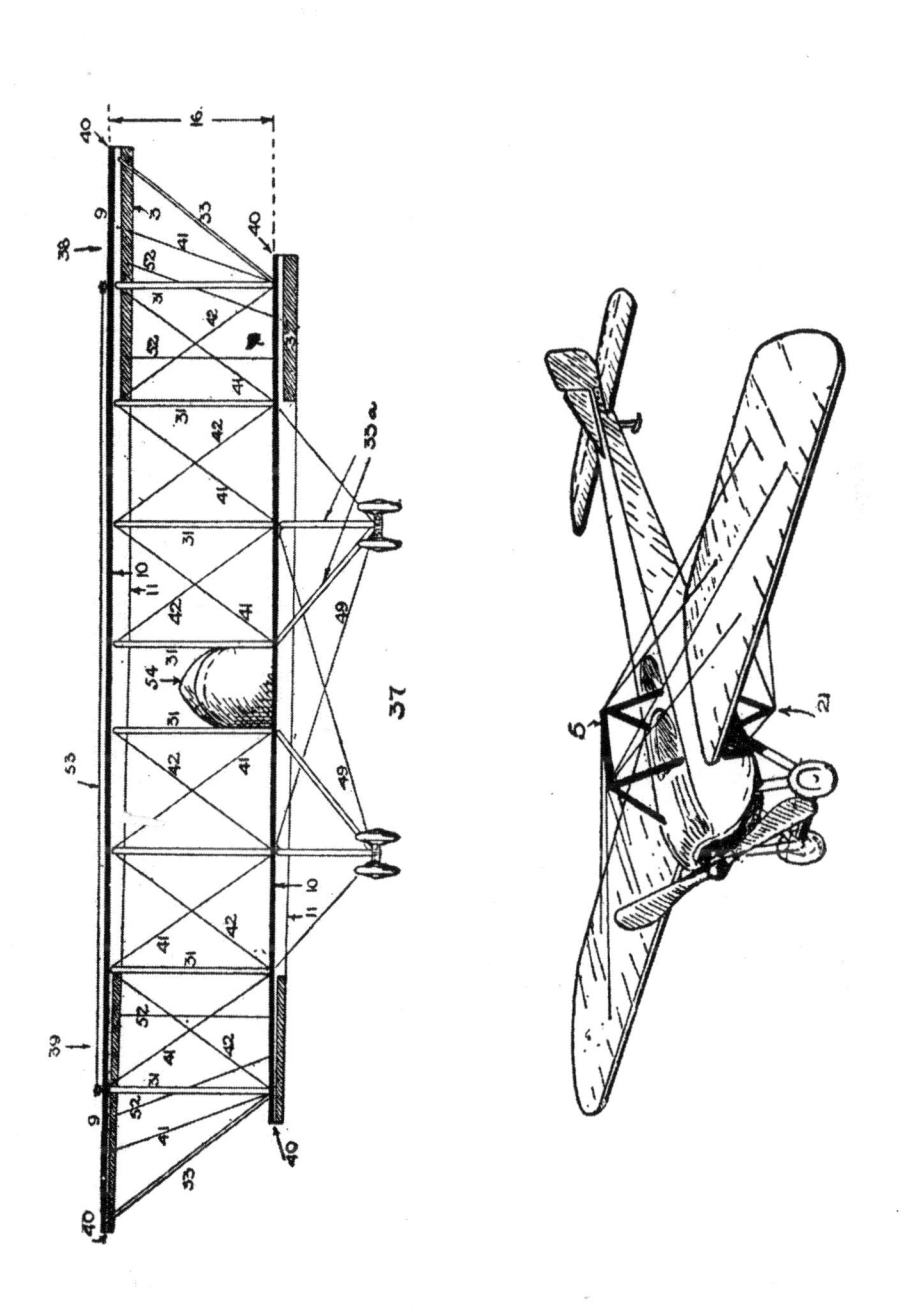

GLOSSARY

The numbers at the right-hand side of the page indicate the parts numbered in the preceding diagrams.

Aeronautics—The science of aerial navigation.

Aerofoil—A rigid structure, of large superficial area relative to its thickness, designed to obtain, when driven through the air at an angle inclined to the direction of motion, a reaction from the air approximately at right angles to its surface. Always cambered when intended to secure a reaction in one direction only. As the term "aerofoil" is hardly ever used in practical aeronautics, I have, throughout this book, used the term SURFACE, which, while academically incorrect, since it does not indicate thickness, is a term usually used to describe the cambered lifting surfaces, *i.e.*, the "planes" or "wings," and the stabilizers and the controlling aerofoils.

Aerodrome—The name usually applied to a ground used for the practice of aviation. It really means "flying machine," but is never used in that sense nowadays.

Aeroplane—A power-driven aerofoil with stabilizing and controlling surfaces.

Acceleration—The rate of change of velocity.

Angle of Incidence—The angle at which the "neutral lift line" of a surface attacks the air.

Angle of Incidence, Rigger's—The angle the chord of a surface makes with a line parallel to the axis of the propeller.

Angle of Incidence, Maximum—The greatest angle of incidence at which, for a given power, surface (including detrimental surface), and weight, horizontal flight can be maintained.

Angle of Incidence, Minimum—The smallest angle of incidence at which, for a given power, surface (including detrimental surface), and weight, horizontal flight can be maintained.

Angle of Incidence, Best Climbing—That angle of incidence at which an aeroplane ascends quickest. An angle approximately halfway between the maximum and optimum angles.

Angle of Incidence, Optimum—The angle of incidence at which the lift-drift ratio is the highest.

THE AEROPLANE SPEAKS

Angle, Gliding—The angle between the horizontal and the path along which an aeroplane at normal flying speed, but not under engine power, descends in still air.

Angle, Dihedral—The angle between two planes.

Angle, Lateral Dihedral—The lifting surface of an aeroplane is said to be at a lateral dihedral angle when it is inclined upward towards its wing-tips.

Angle, Longitudinal Dihedral—The main surface and tail surface are said to be at a longitudinal dihedral angle when the projections of their neutral lift lines meet and produce an angle above them.

Angle, Rigger's Longitudinal Dihedral—Ditto, but substituting "chords" for "neutral life lines."

Angle, Pitch—The angle at any given point of a propeller, at which the blade is inclined to the direction of motion when the propeller is revolving but the aeroplane stationary.

Altimeter—An instrument used for measuring height.

Air-Speed Indicator—An instrument used for measuring air pressures or velocities. It consequently indicates whether the surface is securing the requisite reaction for flight. Usually calibrated in miles per hour, in which case it indicates the correct number of miles per hour at only one altitude. This is owing to the density of the air decreasing with increase of altitude and necessitating a greater speed through space to secure the same air pressure as would be secured by less speed at a lower altitude. It would be more correct to calibrate it in units of air pressure. [1]

Air Pocket—A local movement or condition of the air causing an aeroplane to drop or lose its correct attitude.

Aspect-Ratio—The proportion of span to chord of a surface.

Air-Screw (Propeller)—A surface so shaped that its rotation about an axis produces a force (thrust) in the direction of its axis. [2]

Aileron—A controlling surface, usually situated at the wing-tip, the operation of which turns an aeroplane about its longitudinal axis; causes an aeroplane to tilt sideways. [3]

Aviation—The art of driving an aeroplane.

Aviator—The driver of an aeroplane.

Barograph—A recording barometer, the charts of which can be calibrated for showing air density or height.

Barometer—An instrument used for indicating the density of air.

Bank, to—To turn an aeroplane about its longitudinal axis (to tilt sideways) when turning to left or right.

Biplane—An aeroplane of which the main lifting surface consists of a surface or pair of wings mounted above another surface or pair of wings.

GLOSSARY

Bay—The space enclosed by two struts and whatever they are fixed to.

Boom—A term usually applied to the long spars joining the tail of a "pusher" aeroplane to its main lifting surface. [4]

Bracing—A system of struts and tie wires to transfer a force from one point to another.

Canard—Literally "duck." The name which was given to a type of aeroplane of which the longitudinal stabilizing surface (*empennage*) was mounted in front of the main lifting surface. Sometimes termed "tail-first" aeroplanes, but such term is erroneous, as in such a design the main lifting surface acts as, and is, the *empennage*.

Cabre—To fly or glide at an excessive angle of incidence; tail down.

Camber—Curvature.

Chord—Usually taken to be a straight line between the trailing and leading edges of a surface.

Cell—The whole of the lower suface, that part of the upper surface directly over it, together with the struts and wires holding them together.

Centre (Line) of Pressure—A line running from wing-tip to wing-tip, and through which all the air forces acting upon the surface may be said to act, or about which they may be said to balance.

Centre (Line) of Pressure, Resultant—A line transverse to the longitudinal axis, and the position of which is the resultant of the centres of pressure of two or more surfaces.

Centre of Gravity—The centre of weight.

Cabane—A combination of two pylons, situated over the fuselage, and from which anti-lift wires are suspended. [5]

Cloche—Literally "bell." Is applied to the bell-shaped construction which forms the lower part of the pilot's control lever in a Bleriot monoplane, and to which the control cables are attached.

Centrifugal Force—Every body which moves in a curved path is urged outwards from the centre of the curve by a force termed "centrifugal."

Control Lever—A lever by means of which the controlling surfaces are operated. It usually operates the ailerons and elevator. The "joy-stick." [6]

Cavitation, Propeller—The tendency to produce a cavity in the air.

Distance Piece—A long, thin piece of wood (sometimes tape) passing through and attached to all the ribs in order to prevent them from rolling over sideways. [7]

Displacement—Change of position.

THE AEROPLANE SPEAKS

Drift (*of an aeroplane as distinct from the propeller*)—The horizontal component of the reaction produced by the action of driving through the air a surface inclined upwards and towards its direction of motion *plus* the horizontal component of the reaction produced by the "detrimental" surface *plus* resistance due to "skin-friction." Sometimes termed "head-resistance."

Drift, Active—Drift produced by the lifting surface.

Drift, Passive—Drift produced by the detrimental surface.

Drift (*of a propeller*)—Analogous to the drift of an aeroplane. It is convenient to include "cavitation" within this term.

Drift, to—To be carried by a current of air; to make leeway.

Dive, to—To descend so steeply as to produce a speed greater than the normal flying speed.

Dope, to—To paint a fabric with a special fluid for the purpose of tightening and protecting it.

Density—Mass of unit volume, for instance, pounds per cubic foot.

Efficiency—$\dfrac{\text{Output}}{\text{Input}}$

Efficiency (*of an aeroplane as distinct from engine and propeller*)—
$\dfrac{\text{Lift and Velocity}}{\text{Thrust } (=\text{aeroplane drift})}$

Efficiency, Engine— $\dfrac{\text{Brake horse-power}}{\text{Indicated horse-power}}$

Efficiency, Propeller— $\dfrac{\text{Thrust horse-power}}{\text{Horse-power received from engine}}$
$(=\text{propeller drift})$

NOTE.—The above terms can, of course, be expressed in foot-pounds. It is then only necessary to divide the upper term by the lower one to find the measure of efficiency.

Elevator—A controlling surface, usually hinged to the rear of the tail-plane, the operation of which turns an aeroplane about an axis which is transverse to the direction of normal horizontal flight. [8]

Empennage—See "Tail-plane."

Energy—Stored work. For instance, a given weight of coal or petroleum stores a given quantity of energy which may be expressed in foot-pounds.

Extension—That part of the upper surface extending beyond the span of the lower surface. [9]

Edge, Leading—The front edge of a surface relative to its normal direction of motion. [10]

Edge, Trailing—The rear edge of a surface relative to its normal direction of motion. [11]

GLOSSARY

Factor of Safety—Usually taken to mean the result found by dividing the stress at which a body will collapse by the maximum stress it will be called upon to bear.

Fineness (*of stream-line*)—The proportion of length to maximum width.

Flying Position—A special position in which an aeroplane must be placed when rigging it or making adjustments. It varies with different types of aeroplanes. Would be more correctly described as "rigging position."

Fuselage—That part of an aeroplane containing the pilot, and to which is fixed the tail-plane. [12]

Fin—Additional keel-surface, usually mounted at the rear of an aeroplane. [13]

Flange (*of a rib*)—That horizontal part of a rib which prevents it from bending sideways. [14]

Flight—The sustenance of a body heavier than air by means of its action upon the air.

Foot-pound—A measure of work representing the weight of 1 lb. raised 1 foot.

Fairing—Usually made of thin sheet aluminum, wood, or a light construction of wood and fabric; and bent round detrimental surface in order to give it a "fair" or "stream-like" shape. [15]

Gravity—Is the force of the Earth's attraction upon a body. It decreases with increase of distance from the Earth. See "Weight."

Gravity, Specific—Density of substance
      Density of water.
Thus, if the density of water is 10 lb. per unit volume, the same unit volume of petrol, if weighing 7 lb., would be said to have a specific gravity of $7/10$, *i. e.*, 0.7.

Gap (*of an aeroplane*)—The distance between the upper and lower surfaces of a biplane. In a triplane or multiplane, the distance between a surface and the one first above it. [16]

Gap, Propeller—The distance, measured in the direction of the thrust, between the spiral courses of the blades.

Girder—A structure designed to resist bending, and to combine lightness and strength.

Gyroscope—A heavy circular wheel revolving at high speed, the effect of which is a tendency to maintain its plane of rotation against disturbing forces.

Hangar—An aeroplane shed.

Head-Resistance—Drift. The resistance of the air to the passage of a body.

Helicopter—An air-screw revolving about a vertical axis, the direction of its thrust being opposed to gravity.

THE AEROPLANE SPEAKS

Horizontal Equivalent—The plan view of a body whatever its attitude may be.

Impulse—A force causing a body to gain or lose momentum.

Inclinometer—A curved form of spirit-level used for indicating the attitude of a body relative to the horizontal.

Instability—An inherent tendency of a body, which, if the body is disturbed, causes it to move into a position as far as possible away from its first position.

Instability, Neutral—An inherent tendency of a body to remain in the position given it by the force of a disturbance, with no tendency to move farther or to return to its first position.

Inertia—The inherent resistance to displacement of a body as distinct from resistance the result of an external force.

Joy-Stick—See "Control Lever."

Keel-Surface—Everything to be seen when viewing an aeroplane from the side of it.

King-Post—A bracing strut; in an aeroplane, usually passing through a surface and attached to the main spar, and from the end or ends of which wires are taken to spar, surface, or other part of the construction in order to prevent distortion. When used in connection with a controlling surface, it usually performs the additional function of a lever, control cables connecting its ends with the pilot's control lever. [17]

Lift—The vertical component of the reaction produced by the action of driving through the air a surface inclined upwards and towards its direction of motion.

Lift, Margin of—The height an aeroplane can gain in a given time and starting from a given altitude.

Lift-Drift Ratio—The proportion of lift to drift.

Loading—The weight carried by an aerofoil. Usually expressed in pounds per square foot of superficial area.

Longeron—The term usually applied to any long spar running lengthways of a fuselage. [18]

Mass—The mass of a body is a measure of the quantity of material in it.

Momentum—The product of the mass and velocity of a body is known as "momentum."

Monoplane—An aeroplane of which the main lifting surface consists of one surface or one pair of wings.

Multiplane—An aeroplane of which the main lifting surface consists of numerous surfaces or pairs of wings mounted one above the other.

GLOSSARY

Montant—Fuselage strut.

Nacelle—That part of an aeroplane containing the engine and pilot and passenger, and to which the tail-plane is not fixed. [19]

Neutral Lift Line—A line taken through a surface in a forward direction relative to its direction of motion, and starting from its trailing edge. If the attitude of the surface is such as to make the said line coincident with the direction of motion, it results in no lift, the reaction then consisting solely of drift. The position of the neutral lift line, *i.e.*, the angle it makes with the chord, varies with differences of camber, and it is found by means of wind-tunnel research.

Newton's Laws of Motion—1. If a body be at rest, it will remain at rest; or, if in motion, it will move uniformly in a straight line until acted upon by some force.

2. The rate of change of the quantity of motion (momentum) is proportional to the force which causes it, and takes place in the direction of the straight line in which the force acts. If a body be acted upon by several forces, it will obey each as though the others did not exist, and this whether the body be at rest or in motion.

3. To every action there is opposed an equal and opposite reaction.

Ornithopter (or Orthopter)—A flapping wing design of aircraft intended to imitate the flight of a bird.

Outrigger—This term is usually applied to the framework connecting the main surface with an elevator placed in advance of it. Sometimes applied to the "tail-boom" framework connecting the tail-plane with the main lifting surface. [20]

Pancake, to—To "stall"

Plane—This term is often applied to a lifting surface. Such application is not quite correct, since "plane" indicates a flat surface, and the lifting surfaces are always cambered.

Propeller—See "Air-Screw."

Propeller, Tractor—An air-screw mounted in front of the main lifting surface.

Propeller, Pusher—An air-screw mounted behind the main lifting surface.

Pusher—An aeroplane of which the propeller is mounted behind the main lifting surface.

Pylon—Any V-shaped construction from the point of which wires are taken.

Power—Rate of working. [21]

Power, Horse—One horse-power represents a force sufficient to raise 33,000 lbs. 1 foot in a minute.

THE AEROPLANE SPEAKS

Power, Indicated Horse—The I.H.P. of an engine is a measure of the rate at which work is done by the pressure upon the piston or pistons, as distinct from the rate at which the engine does work. The latter is usually termed "brake horse-power," since it may be measured by an absorption brake.

Power, Margin of—The available quantity of power above that necessary to maintain horizontal flight at the optimum angle.

Pitot Tube—A form of air-speed indicator consisting of a tube with open end facing the wind, which, combined with a static pressure or suction tube, is used in conjunction with a gauge for measuring air pressures or velocities. (*No. 1 in diagram.*)

Pitch, Propeller—The distance a propeller advances during one revolution supposing the air to be solid.

Pitch, to—To plunge nose-down.

Reaction—A force, equal and opposite to the force of the action producing it.

Rudder—A controlling surface, usually hinged to the tail, the operation of which turns an aeroplane about an axis which is vertical in normal horizontal flight; causes an aeroplane to turn to left or right of the pilot. [22]

Roll, to—To turn about the longitudinal axis.

Rib, Ordinary—A light curved wooden part mounted in a fore and aft direction within a surface. The ordinary ribs give the surface its camber, carry the fabric, and transfer the lift from the fabric to the spars. [23]

Rib, Compression—Acts as an ordinary rib, besides bearing the stress of compression produced by the tension of the internal bracing wires. [24]

Rib, False—A subsidiary rib, usually used to improve the camber of the front part of the surface. [25]

Right and Left Hand—Always used relative to the position of the pilot. When observing an aeroplane from the front of it, the right hand side of it is then on the left hand of the observer.

Remou—A local movement or condition of the air which may cause displacement of an aeroplane.

Rudder-Bar—A control lever moved by the pilot's feet, and operating the rudder. [26]

Surface—See "Aerofoil."

Surface, Detrimental—All exterior parts of an aeroplane including the propeller, but excluding the (aeroplane) lifting and (propeller) thrusting surfaces.

Surface, Controlling—A surface the operation of which turns an aeroplane about one of its axes.

GLOSSARY

Skin-Friction—The friction of the air with roughnesses of surface. A form of drift.

Span—The distance from wing-tip to wing-tip.

Stagger—The distance the upper surface is forward of the lower surface when the axis of the propeller is horizontal.

Stability—The inherent tendency of a body, when disturbed, to return to its normal position.

Stability, Directional—The stability about an axis which is vertical during normal horizontal flight, and without which an aeroplane has no natural tendency to remain upon its course.

Stability, Longitudinal—The stability of an aeroplane about an axis transverse to the direction of normal horizontal flight, and without which it has no tendency to oppose pitching and tossing.

Stability, Lateral—The stability of an aeroplane about its longitudinal axis, and without which it has no tendency to oppose sideways rolling.

Stabilizer—A surface, such as fin or tail-plane, designed to give an aeroplane inherent stability.

Stall, to—To give or allow an aeroplane an angle of incidence greater than the "maximum" angle, the result being a fall in the lift-drift ratio, the lift consequently becoming less than the weight of the aeroplane, which must then fall, *i.e.*, "stall" or "pancake."

Stress—Burden or load.

Strain—Deformation produced by stress.

Side-Slip, to—To fall as a result of an excessive "bank" or "roll."

Skid, to—To be carried sideways by centrifugal force when turning to left or right.

Skid, Undercarriage—A spar, mounted in a fore and aft direction, and to which the wheels of the undercarriage are sometimes attached. Should a wheel give way the skid is then supposed to act like the runner of a sleigh and to support the aeroplane. [28]

Skid, Tail—A piece of wood or other material, orientable, and fitted with shock absorbers, situated under the tail of an aeroplane in order to support it upon the ground and to absorb the shock of alighting. [28a]

Section—Any separate part of the top surface, that part of the bottom surface immediately underneath it, with their struts and wires.

Spar—Any long piece of wood or other material.

Spar, Main—A spar within a surface and to which all the ribs are attached, such spar being the one situated nearest to the centre of pressure. It transfers more than half the lift from the ribs to the bracing. [29]

Spar, Rear—A spar within a surface, and to which all the ribs are attached, such spar being situated at the rear of the centre of pressure and at a greater distance from it than is the main spar. It transfers less than half of the lift from the ribs to the bracing. [30]

Strut—Any wooden member intended to take merely the stress of direct compression.

Strut, Interplane—A strut holding the top and bottom surfaces apart. [31]

Strut, Fuselage—A strut holding the *fuselage longerons* apart. It should be stated whether top, bottom, or side. If side, then it should be stated whether right or left hand. *Montant.* [32]

Strut, Extension—A strut supporting an "extension" when not in flight. It may also prevent the extension from collapsing upwards during flight. [33]

Strut, Undercarriage— [33a]

Strut, Dope—A strut within a surface, so placed as to prevent the tension of the doped fabric from distorting the framework. [34]

Serving—To bind round with wire, cord, or similar material. Usually used in connection with wood joints and wire cable splices.

Slip, Propeller—The pitch less the distance the propeller advances during one revolution.

Stream-Line—A form or shape of detrimental surface designed to produce minimum drift.

Toss, to—To plunge tail-down.

Torque, Propeller—The tendency of a propeller to turn an aeroplane about its longitudinal axis in a direction opposite to that in which the propeller revolves.

Tail-Slide—A fall whereby the tail of an aeroplane leads.

Tractor—An aeroplane of which the propeller is mounted in front of the main lifting surface.

Triplane—An aeroplane of which the main lifting surface consists of three surfaces or pairs of wings mounted one above the other.

Tail-Plane—A horizontal stabilizing surface mounted at some distance behind the main lifting surface. *Empennage.* [36]

Turnbuckle—A form of wire-tightener, consisting of a barrel into each end of which is screwed an eyebolt. Wires are attached to the eyebolts and the required degree of tension is secured by means of rotating the barrel.

Thrust, Propeller—See "Air-Screw."

Undercarriage—That part of an aeroplane beneath the *fuselage* or *nacelle,* and intended to support the aeroplane when at rest, and to absorb the shock of alighting. [37]

GLOSSARY

Velocity—Rate of displacement; speed.

Volplane—A gliding descent.

Weight—Is a measure of the force of the Earth's attraction (gravity) upon a body. The standard unit of weight in this country is 1 lb., and is the force of the Earth's attraction on a piece of platinum called *the standard pound,* deposited with the Board of Trade in London. At the centre of the Earth a body will be attracted with equal force in every direction. It will therefore have no weight, though its mass is unchanged. Gravity, of which weight is a measure, decreases with increase of altitude.

Web (*of a rib*)—That vertical part of a rib which prevents it from bending upwards. [37a]

Warp, to—To distort a surface in order to vary its angle of incidence. To vary the angle of incidence of a controlling surface.

Wash—The disturbance of air produced by the flight of an aeroplane.

Wash-in—An increasing angle of incidence of a surface towards its wing-tip. [38]

Wash-out—A decreasing angle of incidence of a surface towards its wing-tip. [39]

Wing-tip—The right- or left-hand extremity of a surface. [40]

Wire—A wire is, in Aeronautics, always known by the name of its function.

Wire, Lift or Flying—A wire opposed to the direction of lift, and used to prevent a surface from collapsing upward during flight. [41]

Wire, Anti-lift or Landing—A wire opposed to the direction of gravity, and used to sustain a surface when it is at rest. [42]

Wire, Drift—A wire opposed to the direction of drift, and used to prevent a surface from collapsing backwards during flight.

Wire, Anti-drift—A wire opposed to the tension of a drift wire, and used to prevent such tension from distorting the framework. [44]

Wire, Incidence—A wire running from the top of an interplane strut to the bottom of the interplane strut in front of or behind it. It maintains the "stagger" and assists in maintaining the angle of incidence. Sometimes termed "stagger wire." [45]

Wire, Bracing—Any wire holding together the framework of any part of an aeroplane. It is not, however, usually applied to the wires described above unless the function performed includes a function additional to those described above. Thus, a lift wire, while strictly speaking a bracing wire, is not usually described as one unless it performs the additional function of bracing some well-defined part such as the undercarriage. It will then be said to be an "undercarriage bracing lift wire." It might, perhaps, be acting as a drift wire also, in which case it will then be described as an "undercarriage bracing lift-drift wire." It should always be stated whether a bracing wire is (1) top, (2) bottom, (3) cross, or (4) side. If a "side bracing wire," then it should be stated whether right- or left-hand.

Wire, Internal Bracing—A bracing wire (usually drift or anti-drift) within a surface.

Wire, Top Bracing—A bracing wire, approximately horizontal and situated between the top longerons of fuselate, between top tail booms, or at the top of similar construction. [46]

Wire, Bottom Bracing—Ditto, substituting "bottom" for "top." [47]

Wire, Side Bracing—A bracing wire crossing diagonally a side bay of fuselage, tail boom bay, undercarriage side bay or centre-section side bay. This term is not usually used with reference to incidence wires, although they cross diagonally the side bays of the cell. It should be stated whether right- or left-hand. [48]

Wire, Cross Bracing—A bracing wire, the position of which is diagonal from right to left when viewing it from the front of an aeroplane. [49]

Wire, Control Bracing—A wire preventing distortion of a controlling surface. [50]

Wire, Control—A wire connecting a controlling surface with the pilot's control lever, wheel, or rudder-bar. [51]

Wire, Aileron Gap—A wire connecting top and bottom ailerons. [52]

Wire, Aileron Balance—A wire connecting the right- and left-hand top ailerons. Sometimes termed the "aileron compensating wire." [53]

Wire, Snaking—A wire, usually of soft metal, wound spirally or tied round another wire, and attached at each end to the framework. Used to prevent the wire round which it is "snaked" from becoming, in the event of its displacement, entangled with the propeller.

Wire, Locking—A wire used to prevent a turnbuckle barrel or other fitting from losing its adjustment.

Wing—Strictly speaking, a wing is one of the surfaces of an ornithopter. The term is, however, often applied to the lifting surface of an aeroplane when such surface is divided into two parts, one being the left-hand "wing," and the other the right-hand "wing."

Wind-Tunnel—A large tube used for experimenting with surfaces and models, and through which a current of air is made to flow by artificial means.

Work—Force × displacement.

Wind-Screen—A small transparent screen mounted in front of the pilot to protect his face from the air pressure.

Types of Aeroplanes

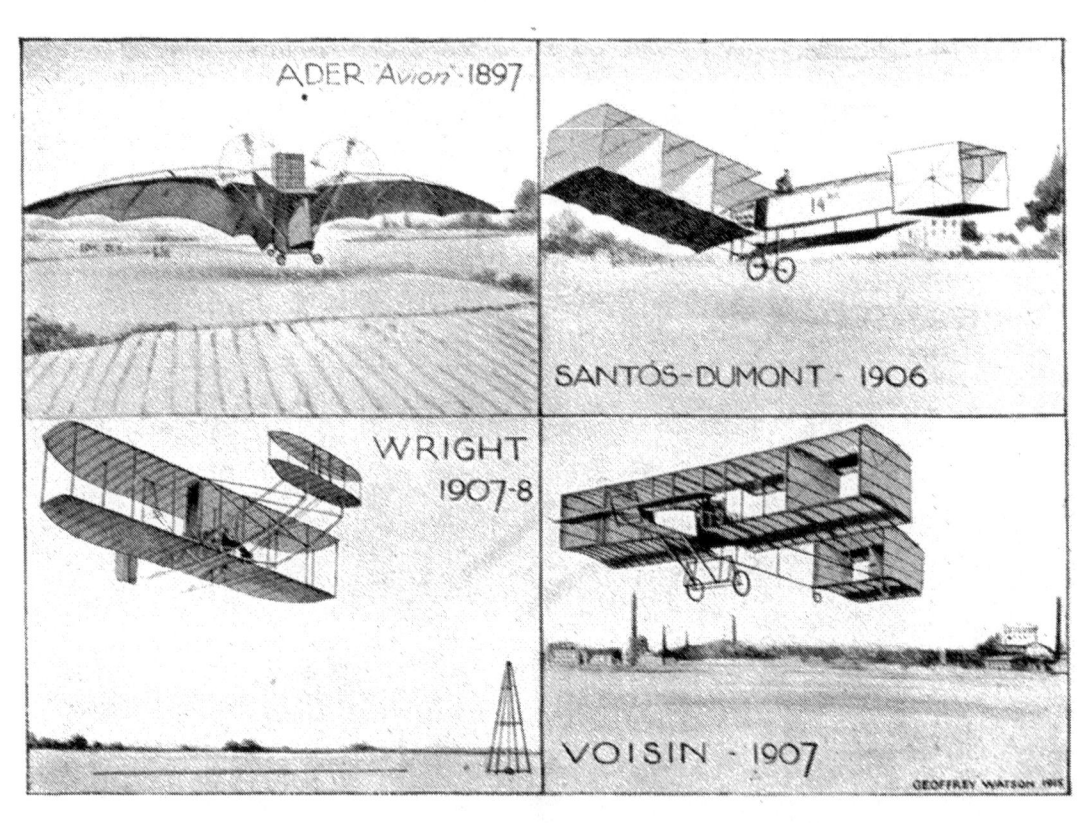

PLATE I.
The first machine to fly—of which there is anything like authentic record—was the Ader "Avion," after which the more notable advances were made as shown above.

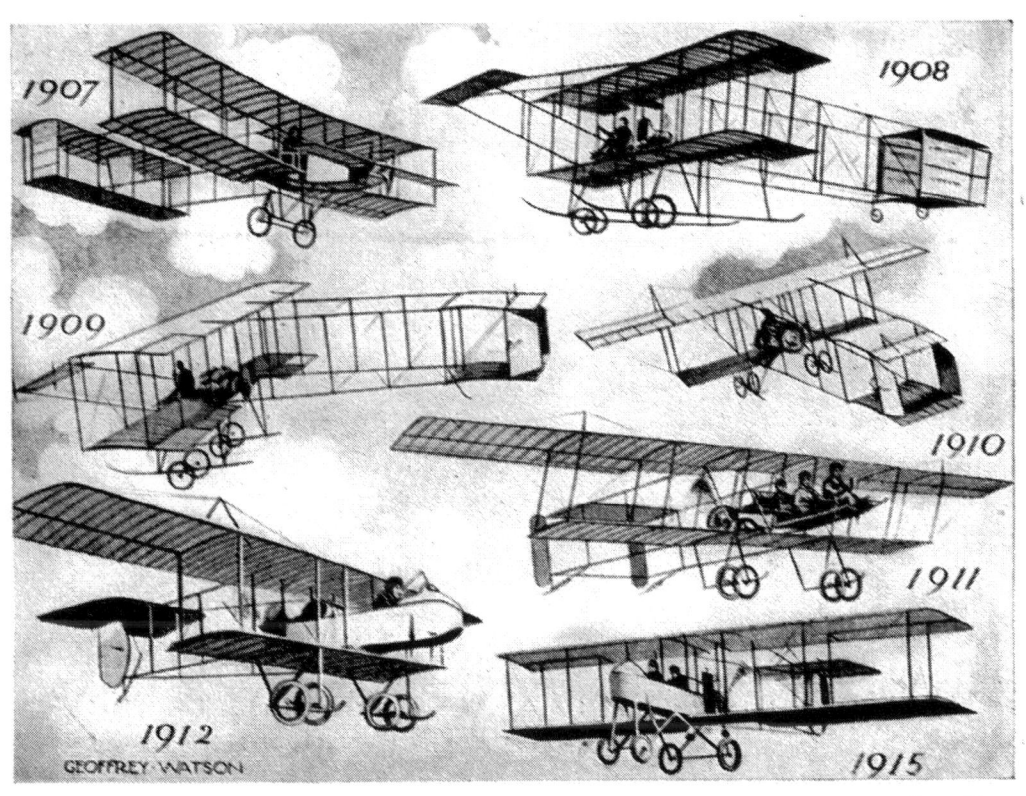

The Henri Farman was the first widely used Aeroplane. Above are shown the chief steps in its development.

PLATE III

THE AVRO.—The aeroplane designed and built by Mr. A. V. Roe was the first successful heavier-than-air flying machine built by a British subject. Mr. Roe's progress may be followed in the picture, from his early "canard" biplane, through various triplanes, with 35 J.A.P. and 35 h.p. Green engines, to his successful tractor biplane with the same 35 h.p. Green, thence through the "totally enclosed" biplane 1912, with 60 h.p. Green, to the biplane 1913-14, with 80 h.p. Gnome.

PLATE IV

THE SOPWITH LAND-GOING BIPLANES.—The earliest was a pair of Wright planes with a fuselage added. Next was the famous tractor with 80 h.p. Gnome. Then the "tabloid" of 1913, which set a completely new fashion in aeroplane design. From this developed the Gordon-Bennett racer shown over date 1914. The gun-carrier was produced about the same time, and the later tractor biplane in a development of the famous 80 h.p. but with 100 h.p. monosoupape Gnome.

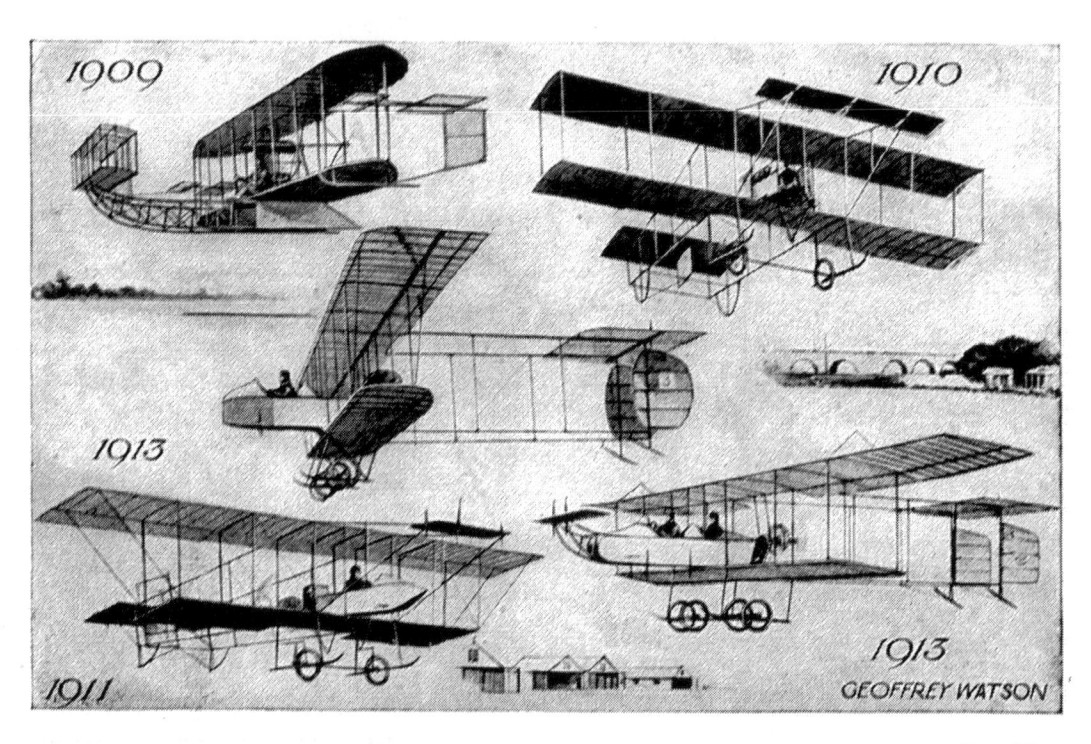

PLATE VI.

THE SHORT "PUSHERS."—In 1909 came the semi-Wright biplane, with 35 h.p. Green, on which Mr. Moore-Brabazon won the "Daily Mail's" £1000 prize for the first mile flight on a circuit on a British aeroplane. Then the first box-kite flown by Mr. Grace at Wolverhampton. Later the famous "extension" type on which the first Naval officers learned to fly. Then the "38" type with elevator on the nacelle, on which dozens of R.N.A.S. pilots were taught.

PLATE VII.

SHORT TRACTORS, 1911-1912.— They were all co-existent, but the first was the "tractor-pusher" (bottom of picture). Then came the "twin-tractor plus propeller" (at top). A development was the "triple-tractor" (on the right), with two 50 h.p. Gnomes, one immediately behind the other under the cowl, one driving the two chains, the other coupled direct. Later came the single-engined 80 h.p. tractor (on the left), the original of the famous Short seaplanes.

PLATE VIII.

THE VICKERS MACHINES: First the Vickers-R.E.P. of 1911, which developed into the full-bodied No. V. with R.E.P. engine, then the Military Trials "sociable" with Viale engine, and so to the big No. VII with a 100 h.p. Gnome. Contemporary with the No. V and No. VI were a number of school box-kites of ordinary Farman type, which developed into the curious "pumpkin" sociable, and the early "gun 'bus" of 1913. Thence arrived the gun-carrier with 100 h.p. monosoupape Gnome.

PLATE IX.

THE BRISTOL AEROPLANES.—First, 1910, Farman type box-kites familiar to all early pupils. Then the miniature Maurice-Farman type biplane of the "Circuit of Britain." Contemporaneous was the "floating tail" monoplane designed by Pierre Prier, and after it a similar machine with fixed tail. Then came the handsome but unfortunate monoplane designed by M. Coanda for the Military Trials, 1912.

PLATE X.

THE BRISTOL TRACTORS.—Late 1912 came the round fuselaged tractor, with Gnome engine, designed by Mr. Gordon England for Turkey. 1912-13 came the biplane built onto the Military Trials monoplane type fuselage, also with a Gnome, designed by M. Coanda for Roumania. Then the Renault-engined Coanda tractor 1913, followed by 80 h.p. Gnome-engined scout, designed by Messrs. Barnwell and Busteed, which with Gnomes, le Rhones and Clergets, has been one of the great successes. Almost contemporary was the two-seater Bristol.

PLATE XI.

THE MARTINSYDES.—1909, first experimental monoplane built with small 4-cylinder engine. J.A.P.-engined machine, 1910, followed by the Gnome-engined machine, 1911. 1912, first big monoplane with Antoinette engine was built, followed by powerful Austro-Daimler monoplane, 1913. Then came the little Gnome-engined scout biplanes, 1914, some with, some without, skids.

PLATE XII.

THE CURTISS BIPLANES.—In 1909 came the "June-bug," the united product of Glen Curtiss, Dr. Graham Bell, and J. A. D. McCurdy. Then the box-kite type, 1909, on which Mr. Curtiss won the Gordon-Bennett Race at Reims. Next the "rear-elevator" pusher, 1912, followed by first tractor, 1913, with an outside flywheel. All purely Curtiss machines to that date had independent ailerons intended to get away from Wright patents. Following these came tractors with engines varying from 70 to 160 h.p., fitted with varying types of chassis. All these have ordinary ailerons.

PLATE XIII.

THE BLERIOT (1).—The first engine-driven machine was a "canard" monoplane. Then came the curious tractor monoplanes 1908-1909, in order shown. Famous "Type XI" was prototype of all Bleriot successes. "Type XII" was never a great success, though the ancestor of the popular "parasol" type. The big passenger carrier was a descendant of this type.

PLATE XIV

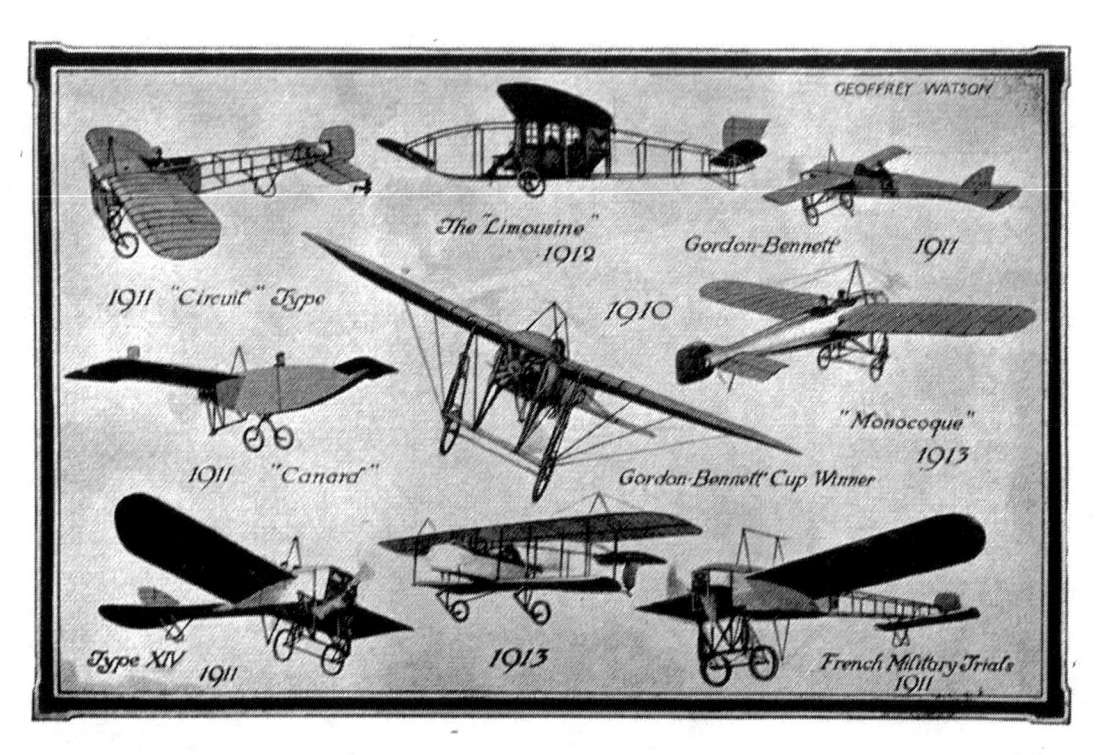

THE BLERIOT (2):—1910, "Type XI," on which Mr. Grahame-White won Gordon-Bennett Race, with a 14-cylinder 100 h.p. Gnome. 1911 came the improved "Type XI," with large and effective elevator flaps. On this type, with a 50 h.p. Gnome, Lieut. de Conneau (M. Beaumont) won Paris-Rome Race and "Circuit of Britain." Same year saw experimental "Limousine" flown by M. Legagneux, and fast but dangerous "clipped-wing" Gordon-Bennett racer with the fish-tail, flown by Mr. Hamel. About the same time came the fish-tailed side-by-side two-seater, flown by Mr. Hamel at Hendon and by M. Perreyon in 1912 Military Trials. 1911, M. Bleriot produced the 100 h.p. three-seater which killed M. Desparmets in French Military Trials. 1912-13, M. Bleriot produced a quite promising experimental biplane, and a "monocoque" monoplane in which the passenger faced rearward.

PLATE XV

THE BLERIOT (3)—1912 tandem two-seater proved one of the best machines of its day. 1913 "canard" lived up to its name. A "pusher" monoplane was built in which the propeller revolved on the top tail boom. This machine came to an untimely end, with the famous pilot, M. Perreyon. 1912 "tandem" was developed in 1914 into the type shown in centre; almost simultaneously "parasol" tandem appeared. 1914, M. Bleriot built a monoplane embodying a most valuable idea never fully developed. The engine tanks and pilot were all inside an armoured casing. Behind them the fuselage was a "monocoque" of three-ply wood bolted onto the armour. And behind this all the tail surfaces were bolted on as a separate unit.

PLATE XVI.

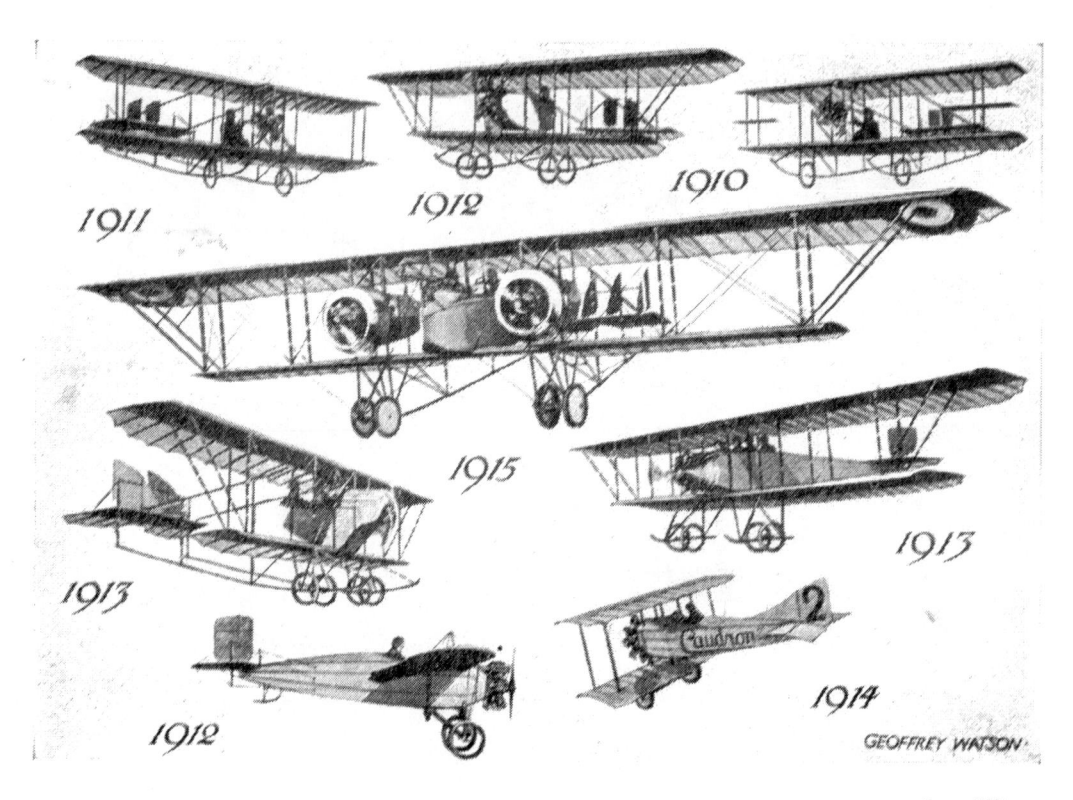

THE CAUDRON.—1910, came the machine with ailerons and a 28 h.p. Anzani. 1911 this was altered to warp control and a "star" Anzani was fitted. From this came the 35 h.p. type of 1912, one of the most successful of school machines. Small fast monoplane, 1912 was never further developed. 1913 appeared the familiar biplanes with 80 h.p. Gnomes, and 5-seater with 100 h.p. Anzani for French "Circuit of Anjou." 1914 produced the "scout" biplane which won at Vienna 1915 appeared the twin-engined type, the first successful "battle-plane."

PLATE XVII.
THE DEPERDUSSIN.—In 1911 the little monoplane with a Gyp. engine. Then the Gnome-engined machine of the "Circuit of Europe." In 1912 came the Navy's machine with 70-h.p. Gnome, and Prevost's Gordon-Bennett "Bullet," 135 miles in the hour. The last was the British-built "Thunder-Bug," familiar at Hendon.

PLATE XVIII

THE BREGUET.—First to fly was the complicated but business-like machine of 1909. Then came the record passenger carrier, 1910 (which lifted 8 passengers). 1911 the French Military Trials machine with geared-down 100 h.p. Gnome appeared. 1912 produced the machine with 130 h.p. Salmson engine on which the late Mr. Moorhouse flew the Channel with Mrs. Moorhouse and Mr. Ledéboer as passengers; also the machine with 130 h.p. horizontal Salmson, known as the "Whitebait." The last before the war was the rigid wing machine with 200 h.p. Salmson.

PLATE XIX.

THE CODY.--First the Military Experiment of 1908, with an Antoinette engine, then improved type 1909 with a Green engine. Next the "Cathedral," 1910, with a Green engine, which won Michelin Prize. In 1911 "Daily Mail" Circuit machine, also with a Green, won the Michelin. This was modified into 1912 type which won Military Competition and £5,000 in prizes, with an Austro-Daimler engine, and later the Michelin Circuit Prize, again with a Green. 1912 the only Cody Monoplane was built. 1913 a modified biplane on which the great pioneer was killed.

PLATE XX.

THE NIEUPORT.—The first Nieuport of 1909 was curiously like a monoplane version of a Caudron. In 1910 came the little two-cylinder machine with fixed tail-plane and universally jointed tail. In 1911 the French Trials machine was built with 100 h.p. 14 cylinder Gnome, and is typical of this make. Also the little two-cylinder record breaker. A modification of 1913 was the height record machine of the late M. Legagneux.

PLATE XXI

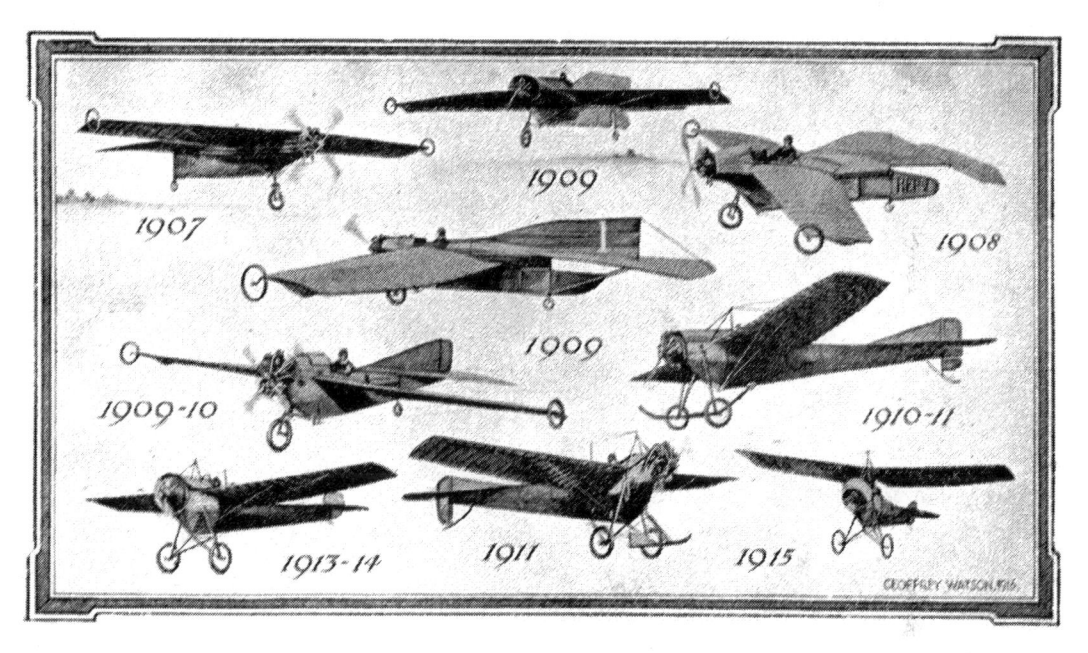

THE R.E.P. MONOPLANES.—First came the curious and highly interesting experiments of 1907, 1908, 1909, and 1910. 1910-1911, the World's Distance Record breaker was produced; after it, the "European Circuit," all with R.E.P. engines. In 1913-14 came the French military type with Gnome engine and finally the "parasol," 1915.

GEOFFREY WATSON

PLATE XXII.

THE MORANE: First the European Circuit and Paris-Madrid type. Then the 1912 types, with taper wing and modern type wing. The 1913 types, the "clipped wing," flown by the late Mr. Hamel, one of the standard tandem types now in use. About the same time came the "parasol." 1914-15 came a little biplane like a Nieuport, and the "destroyer" type with a round section body, flown by Vedrines.

PLATE XXIII.

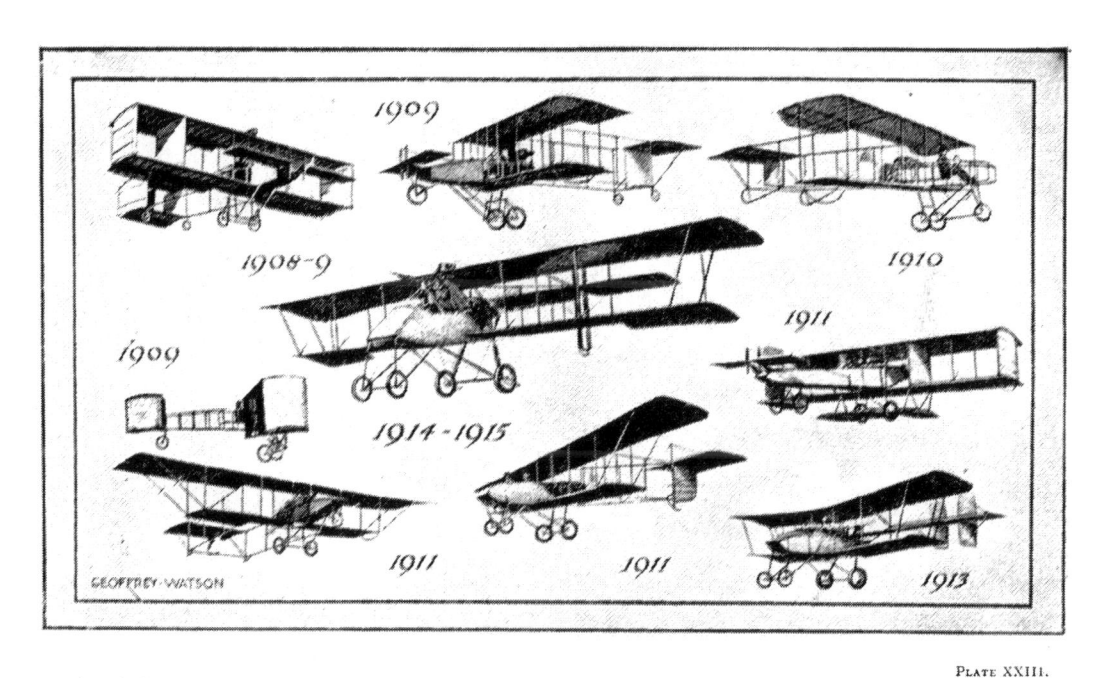

THE VOISIN.—1908, the first properly controlled flight on a European aeroplane was made on a Voisin of the type shown with fixed engine. Then followed the record breaker of 1909 with a Gnome engine. In 1909 also the only Voisin tractor was produced. 1910 the Paris-Bordeaux type was built; 1911 the amphibious "canard" and the "military" type with extensions, and the type without an elevator. 1913 came the type with only two tail-booms and a geared-down engine, which developed into the big "gun" machine with a Salmson engine.

PLATE XXIV.

THE HANRIOT AND PONNIER MONOPLANES.—In 1909 came the first Hanriot with 50 h.p. 6-cylinder Buchet engine, and in 1910 the famous "Henrietta" type with E.N.Vs. and stationary Clergets. 1911 came the Clerget two-seater entered in French Military Trials, and 1912 the 100 h.p. Hanriot-Pagny monoplane which took part in British Military Trials. Sister machines of the same year were the single seater with 50 h.p. Gnome and the 100 h.p. Gnome racer with stripped chassis. In 1913 the Ponnier-Pagny racing monoplane with 160 h.p. Le Rhone competed in the Gordon-Bennett race, doing about 130 miles in the hour. The 60 h.p. Ponnier biplane was the first successful French scout tractor biplane.

PLATE XXV.

THE WRIGHT BIPLANE.—The first power flights were made, 1903, on a converted glider fitted with 16 h.p. motor. The **prone** position of the pilot will be noted. By 1907 the machine had become reasonably practical with 40 h.p. motor. On this **the first** real flying in the world was done. In 1910 the miniature racing Wright was produced; also the type with a rear **elevator** in addition to one in front. Soon afterwards the front elevator disappeared, and the machine became the standard American exhibition and school machine for four years. In 1915 a machine with enclosed fuselage was produced.

PLATE XXVI.

THE BLACKBURN MONOPLANES.—In 1909 was built the curious four-wheeled parasol-type machine with 35 h.p. Green engine and chain transmission, on which flying was done at Saltburn. In 1911 the Isaacson-engined machine was built, together with a 50 h.p. Gnome single-seater on which Mr. Hucks started in the Circuit of Britain race. In 1912 another 50 h.p. single-seater was built on which a good deal of school work was done. A more advanced machine appeared in 1913 and a two-seater with 80 h.p. Gnome did a great deal of cross-country work in 1913-14.

PLATE XXVII.

In 1908 the first Antoinette monoplane was produced by MM. Gastambide and Mengin. Then followed a machine with central skids a single wheel and wing s ids. In 1909 came the machine with four-wheeled chassis and ailerons and later an improved edit on which reverted to the cent al skid idea. On this M. Latham made his first cross-channel att mpt. The next machine shed the wing skids and widened its wheelbase. During 1910-11 the ailerons vanished, warp control was adopted and the king-post system of wing-bracing was used. In 1911 the curious machine with streamlined "pantalette" chassis, totally enclosed body and internal w ng-bracing, was produced for French Military Trials. In 1912 the three-wheeled machine was used to a certain extent in the French Army. Then the type disappeared.

177

PLATE XXVIII.

In 1908 and 1909 detached experimental machines in various countries attained a certain success. The late Capt. Ferber made a primitive tractor biplane 1908. The Odier-Vendome biplane was a curious bat-winged pusher biplane built 1909. The tailless Etrich monoplane, built in Austria, 1908, was an adaptation of the Zanonia leaf. M. Santos-Dumont made primitive parasol type monoplanes known as "Demoiselles," in which bamboo was largely used. 1909 type is seen above. A curious steel monoplane was built by the late John Moisant, 1909. The twin-pusher biplane, built by the Barnwell Bros. in Scotland made one or two straight flights in 1909. The Clement-Bayard Co. in France constructed in 1909 a biplane which did fairly well. Hans Grade, the first German to fly, made his early efforts on a "Demoiselle" type machine, 1908.

PLATE XXIX.

In 1910 a number o novel machines were produced. The Avis with Anzani engine was flown by the Hon. Alan Boyle. Note the cruciform universally jointed tail. The Goupy with 50 h.p. Gnome was an early French tractor, notable for its hinging wing-tips. The Farman was a curious "knock-up" job, chiefly composed of standard box-kite fittings. The Sommer with 50 h.p. Gnome was a development of the box-kite with a shock-breaking chassis. The Savary, also French, was one of the first twin tractors to fly The model illustrated had an E.N.V. engine. Note position of the rudders on the wing tips. The Austr an Etrich was the first successful machine of the Taube class ever built.

PLATE XXX.

INTERESTING MACHINES, 1910.—The Werner monoplane with E.N.V. engine, combined shaft and chain drive, was a variant of the de Pischoff. The Macfie biplane was a conventional biplane with 50 h.p. Gnome and useful originalities. The Valkyrie monoplane, another British machine, was a "canard" monoplane with propeller behind the pilot and in front of main plane. The Weiss monoplane was a good British effort at inherent stability. The Tellier monoplane was a modified Bleriot with Antoinette proportions. The Howard Wright biplane was a pusher with large lifting monoplane tail. The Dunne biplane was another British attempt at inherent stability. The Jezzi biplane was an amateur built twin-propeller.

PLATE XXXI.

SOME INTERESTING MACHINES, 1911.—The Compton-Paterson biplane was very similar to the early Curtiss pusher; it had a 50 b.p. Gnome. The Sloan bicurve was a French attempt at inherent stability with 50 h.p. Gnome and tractor screw. The Paulhan biplane was an attempt at a machine for military purposes to fold up readily for transport. The Sanders was a British biplane intended for ro gh service. The Barnwell monoplane was the first Scottish machine to fly; it had a horizontally opposed Scottish engine. The Harlan monoplane was an early German effort; note position of petrol tank.

PLATE XXXII.

The Clement-Bayard monoplane, 1911, was convertible into a tractor biplane. The standard engine was a 50 h.p. Gnome. The machine was interesting, but never did much. The Zodiac was one of the earliest to employ staggered wings. With 50 h.p. Gnome engine it was badly underpowered, so never did itself justice. The Jezzi tractor biplane, 1911, was a development of an earlier model built entirely by Mr. Jezzi, an amateur constructor. With a low-powered J.A.P. engine it developed an amazing turn of speed, and it may be regarded as a forerunner of the scout type and the properly streamlined aeroplane. The Paulhan-Tatin monoplane, 1911, was a brilliant attempt at high speed for low power; it presented certain advantages as a scout. A 50 h.p. Gnome, fitted behind the pilot's seat in the streamlined fuselage, was cooled through louvres. The propeller at the end of the tail was connected with the engine by a flexible coupling. This machine was, in its day, the fastest for its power in the world, doing 80 miles per hour. Viking I was a twin tractor biplane driven by a 50 h.p. Gnome engine through chains. It was built by the author at Hendon in 1912.

PLATE XXXIII.

Much ingenuity was exerted by the French designers in 1911 to produce machines for the Military Trials. Among them was the 100 h.p. Gnome-Borel monoplane with a four-wheeled chassis, and the Astra triplane with a 75 h.p. Renault engine. This last had a surface of about 500 square feet and presented considerable possibilities. Its principal feature was its enormous wheels with large size tyres as an attempt to solve difficulties of the severe landing tests. The Clement-Bayard biplane was a further development of the Clement-Bayard monoplane; the type represented could be converted into a monoplane at will. The Lohner Arrow biplane with the Daimler engine was an early German tractor biplane built with a view to inherent stability, and proved very successful. The Pivot monoplane was of somewhat unconventional French construction, chiefly notable for the special spring chassis and pivoted ailerons at the main planes; this pivoting had nothing to do with the name of the machine, which was designed by M. Pivot.

PLATE XXXIV.

The Flanders monoplane, 1912, with 70 h.p. Renault engine, was one of the last fitted with king-post system of wing bracing. The Flanders biplane entered for British Military Trials. Notable features: the highly staggered planes, extremely low chassis and deep fuselage. Also, the upper plane was bigger in every dimension than the lower; about the first instance of this practice. The Bristol biplane, with 100 h.p. Gnome engine, was also entered for the Trials, but ultimately withdrawn. The Mars monoplane, later known as the "D.F.W.," was a successful machine of Taube type with 120 h.p. Austro-Daimler engine. The building of the engine into a cowl, complete with radiator in front, followed car practice very closely. The tail of the monoplane had a flexible trailing edge; its angle of incidence could be varied from the pilot's seat, so that perfect longitudinal balance was attained at all loadings and speeds. The Handley-Page monoplane, with 70 h.p. Gnome engine, was an early successful British attempt at inherent stability.

PLATE XXXV.

The Sommer monoplane, with 50 h.p. Gnome, was a 1911-12 machine; it did a good deal of cross-country flying. The Vendome monoplane of 1912, also with 50 h.p. Gnome engine was notable chiefly for its large wheels and jointed fuselage, which enabled the machine to be taken down for transport. The Savary biplane took part in the French Military Trials, 1911. It had a four-cylinder Lator aviation motor. Notable features are twin chain-driven propellers, rudders between the main planes, the broad wheel-base and the position of the pilot. The Paulhan triplane, which also figured in the French Military Trials, was a development of the Paulhan folding biplane. It had a 70 h.p. Renault engine. For practical purposes it was a failure. The R.E.P. biplane, with 60 h.p. R.E.P. engine, was a development of the famous R.E.P. monoplanes. Its spring chassis, with sliding joints, marked an advance. Like the monoplanes, it was built largely of steel.

1911

PLATE XXXVI.

In 1912 came the first really successful Handley Page monoplane, with 50 h.p Gnome engine. The Short monoplane, was built generally on Bleriot lines. Its chassis was an original feature. The Coventry Ordnance biplane was a two-seater tractor built for the British Military Trials. It had a 100 h.p. 14-cylinder Gnome engine, with propeller geared down through a chain drive. The machine was an interesting experiment, but not an unqualified success. The Moreau "Aerostable," fitted with a 50 h.p. Gnome, was a French attempt to obtain automatic stability, but it only operated longitudinally. The pilot's nacelle was pivoted under the main planes, wires were attached to the control members so that the movements of the nacelle in its efforts to keep a level keel brought them into operation. The Mersey monoplane, an entrant for the British Military Trials, was designed to present a clear field of view and fire. The 45 h.p. Isaacson engine was connected by a shaft to a propeller mounted behind the nacelle on the top tail boom. It was a promising experiment, but came to grief. The Radley-Moorhouse monoplane was a sporting type machine on Bleriot lines, with 50 h.p. Gnome engine. It was notable for its streamline body and disc wheels.

More History from CGR Publishing
www.CGRpublishing.com

THE ART OF WORLD WAR I

BY EPHRAIM DURNST

Other books from CGR Publishing at Amazon.com

1939 New York World's Fair: The World of Tomorrow in Photographs

San Francisco 1915 World's Fair: The Panama-Pacific International Expo.

1904 St. Louis World's Fair: The Louisiana Purchase Exposition in Photographs

Chicago 1933 World's Fair: A Century of Progress in Photographs

19th Century New York: A Dramatic Collection of Images

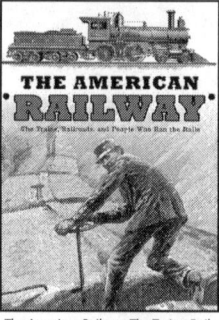

The American Railway: The Trains, Railroads, and People Who Ran the Rails

The Story of the Ship

The World's Fair of 1893 Ultra Massive Photographic Adventure Vol. 1

The World's Fair of 1893 Ultra Massive Photographic Adventure Vol. 2

The World's Fair of 1893 Ultra Massive Photographic Adventure Vol. 3

World War 1: A Dramatic Collection of Images

The White City of Color: 1893 World's Fair

Ethel the Cyborg Ninja Book 1

Ethel they Cyborg Ninja 2

How To Draw Digital by Mark Bussler

How To Draw Pandas by Mark Bussler

Other books from CGR Publishing at Amazon.com

- Ultra Massive Video Game Console Guide Volume 1
- Ultra Massive Video Game Console Guide Volume 2 
- Ultra Massive Video Game Console Guide Volume 3 
- Ultra Massive Sega Genesis Guide 
- Discovery of the North Pole: The Greatest American Expedition 
- Chicago's White City Cookbook 
- Official Guide to the World's Columbian Exposition 
- How To Grow Mushrooms: A 19th Century Approach 
- The Great War Remastered WW1 Standard History Collection Vol. 1 
- Sinking of the Titanic: The Greatest Disaster at Sea 
- All Hail the Vectrex Ultimate Collector's Guide 
- Old Timey Pictures with Silly Captions Volume 1 
- The Art of World War 1 
- The Kaiser As I Know Him: As Told by his Dentist 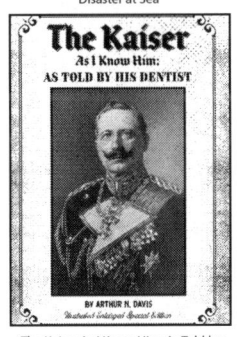
- Captain William Kidd and the Pirates and Buccaneers Who Ravaged the Seas 
- Robot Kitten Factory Issue #1

The Aeroplane Speaks:
Illustrated Historical Guide to Airplanes
Enlarged Special Edition

Copyright © 2020 Inecom, LLC.
All Rights Reserved

www.CGRpublishing.com